Five Years of Separations

Five Years of Separations

Feature Paper 2018

Editors

Victoria Samanidou
Rafael Lucena

MDPI • Basel • Beijing • Wuhan • Barcelona • Belgrade • Manchester • Tokyo • Cluj • Tianjin

Editors
Victoria Samanidou
Laboratory of Analytical Chemistry,
School of Chemistry,
Aristotle University of Thessaloniki
Greece

Rafael Lucena
Department of Analytical Chemistry,
Universidad de Cordoba
Spain

Editorial Office
MDPI
St. Alban-Anlage 66
4052 Basel, Switzerland

This is a reprint of articles from the Special Issue published online in the open access journal *Separations* (ISSN 2297-8739) (available at: https://www.mdpi.com/journal/separations/special_issues/five_years_separations).

For citation purposes, cite each article independently as indicated on the article page online and as indicated below:

LastName, A.A.; LastName, B.B.; LastName, C.C. Article Title. *Journal Name* **Year**, *Article Number*, Page Range.

ISBN 978-3-03936-924-9 (Hbk)
ISBN 978-3-03936-925-6 (PDF)

Cover image courtesy of Victoria Samanidou.

© 2020 by the authors. Articles in this book are Open Access and distributed under the Creative Commons Attribution (CC BY) license, which allows users to download, copy and build upon published articles, as long as the author and publisher are properly credited, which ensures maximum dissemination and a wider impact of our publications.

The book as a whole is distributed by MDPI under the terms and conditions of the Creative Commons license CC BY-NC-ND.

Contents

About the Editors vii

Preface to "Five Years of Separations" ix

Natalia Manousi and Victoria F. Samanidou
Applications of Gas Chromatography for the Analysis of Tricyclic Antidepressants in Biological Matrices
Reprinted from: *Separations* 2019, 6, 24, doi:10.3390/separations6020024 1

Irene Panderi, Eugenia Taxiarchi, Constantinos Pistos, Eleni Kalogria and Ariadni Vonaparti
Insights into the Mechanism of Separation of Bisphosphonates by Zwitterionic Hydrophilic Interaction Liquid Chromatography: Application to the Quantitation of Risedronate in Pharmaceuticals
Reprinted from: *Separations* 2019, 6, 6, doi:10.3390/separations6010006 21

Maria Celeiro, Lua Vazquez, J. Pablo Lamas, Marlene Vila, Carmen Garcia-Jares and Maria Llompart
Miniaturized Matrix Solid-Phase Dispersion for the Analysis of Ultraviolet Filters and Other Cosmetic Ingredients in Personal Care Products
Reprinted from: *Separations* 2019, 6, 30, doi:10.3390/separations6020030 35

Anastasios I. Zouboulis, Efrosyni N. Peleka and Anastasia Ntolia
Treatment of Tannery Wastewater with Vibratory Shear-Enhanced Processing Membrane Filtration
Reprinted from: *Separations* 2019, 6, 20, doi:10.3390/separations6020020 49

Samantha L. Bowerbank, Michelle G. Carlin and John R. Dean
Dissolution Testing of Single- and Dual-Component Thyroid Hormone Supplements
Reprinted from: *Separations* 2019, 6, 18, doi:10.3390/separations6010018 67

Ana Isabel Argente-García, Lusine Hakobyan, Carmen Guillem and Pilar Campíns-Falcó
Estimating Diphenylamine in Gunshot Residues from a New Tool for Identifying both Inorganic and Organic Residues in the Same Sample
Reprinted from: *Separations* 2019, 6, 16, doi:10.3390/separations6010016 75

Ewa Skoczyńska and Jacob de Boer
Retention Behaviour of Alkylated and Non-Alkylated Polycyclic Aromatic Hydrocarbons on Different Types of Stationary Phases in Gas Chromatography
Reprinted from: *Separations* 2019, 6, 7, doi:10.3390/separations6010007 95

Cheryl Frankfater, Robert B. Abramovitch, Georgiana E. Purdy, John Turk, Laurent Legentil, Loïc Lemiègre and Fong-Fu Hsu
Multiple-stage Precursor Ion Separation and High Resolution Mass Spectrometry toward Structural Characterization of 2,3-Diacyltrehalose Family from *Mycobacterium tuberculosis*
Reprinted from: *Separations* 2019, 6, 4, doi:10.3390/separations6010004 107

Filip S. Ekholm, Suvi-Katriina Ruokonen, Marina Redón, Virve Pitkänen, Anja Vilkman, Juhani Saarinen, Jari Helin, Tero Satomaa and Susanne K. Wiedmer
Hydrophilic Monomethyl Auristatin E Derivatives as Novel Candidates for the Design of Antibody-Drug Conjugates
Reprinted from: *Separations* 2019, 6, 1, doi:10.3390/separations6010001 121

Naysla Paulo Reinert, Camila M. S. Vieira, Cristian Berto da Silveira, Dilma Budziak and Eduardo Carasek
A Low-Cost Approach Using Diatomaceous Earth Biosorbent as Alternative SPME Coating for the Determination of PAHs in Water Samples by GC-MS
Reprinted from: *Separations* **2018**, *5*, 55, doi:10.3390/separations5040055 . . . . **133**

About the Editors

Victoria Samanidou was born on 11 January, 1963, in Thessaloniki, Greece. She obtained her Bachelor of Science degree in Chemistry in 1985 from the Chemistry Department of Aristotle University of Thessaloniki, Greece.

In 1990, she obtained a doctorate (Ph.D.) in Chemistry from the Department of Chemistry of the Aristotle University of Thessaloniki. The topic of her thesis was "Distribution and mobilization of heavy metals in waters and sediments from rivers in Northern Greece". In the same year, Dr. Samanidou joined the Laboratory of Analytical Chemistry, in the Department of Chemistry of Aristotle University of Thessaloniki, as a Technical Assistant. Nine years later, she was elected as Lecturer in the Laboratory of Analytical Chemistry in the Department of Chemistry of the Aristotle University of Thessaloniki. In 2007, she joined the Institute of Analytical Chemistry and Radiochemistry in Graz Technical University, for four months, developing methods by LC-MS/MS. Since 2015, Dr. Samanidou has been Full Professor in the Laboratory of Analytical Chemistry in the Department of Chemistry of Aristotle University of Thessaloniki, Greece, where she currently serves as Director of the Laboratory.

Dr. Samanidou has authored and co-authored more than 170 original research articles in peer-reviewed journals and 45 reviews and 50 chapters in scientific books, with an H-index 36 (Scopus June 2020, http://orcid.org/0000-0002-8493-1106, Scopus Author ID 7003896015) and circa 3500 citations. She has supervised four PhD Theses, 24 postgraduate Diploma Theses, 2 postdoc researchers, and more than 15 undergraduate Diploma Theses. She has served as Member of 10 advisory PhD committees, 21 examination PhD committees, and 32 examination committees of postgraduate Diploma Theses. She is a member of the editorial board of more than 10 scientific journals, and she has reviewed circa 500 manuscripts in more than 100 scientific journals. She was also guest editor in more than 10 special issues in scientific journals. She has served as Academic Editor for Separations MDPI, as Regional editor in *Current Analytical Chemistry*, and as Editor in Chief of *Pharmaceutica Analytica Acta*.

Her research interests include: 1. Development and validation of analytical methods for the determination of inorganic and organic substances using chromatographic techniques; 2. development and optimization of methodology for sample preparation of various samples, e.g., food, biological fluids; 3. study of new chromatographic materials used in separation and sample preparation (polymeric sorbents, monoliths, carbon nanotubes, fused core particles, etc.) compared to conventional materials.

She has also been Member of the organizing and scientific committee in 20 scientific conferences. In December 2015, Dr. Samanidou was elected as President of the Steering Committee of the Division of Central and Western Macedonia of the Greek Chemists' Association. In November 2018, she was reelected to serve at the same leading position for 3 more years. A milestone in her career was in 2016, when she was included in the top 50 power list of women in Analytical Science, as proposed by Texere Publishers. https://theanalyticalscientist.com/power-list/the-power-list-2016.

Rafael Lucena has been a professor of Analytical Chemistry at the University of Córdoba since 2010 and the secretary of the Research Institute of Fine Chemistry and Nanochemistry at the same university. To date, Rafael has co-authored 109 scientific articles (h-index: 34, Scopus ID: 8517312100Scopus) dealing with different analytical chemistry facets, although sample preparation is the central core of his research. His research is mainly focused on the development of new extraction approaches and the design of novel materials for analytes isolation. Since 2017, he has worked on the design of a new paper-based sorptive phase with application in analytical chemistry and catalysis. He is also interested in the integration of microextraction techniques with sampling and instrumental analysis (spectroscopic and mass spectrometry techniques).

Rafael has reviewed more than 500 articles in many scientific journals and has evaluated research projects for different agencies such as Fondecyt (Chile), Agencia Estatal de Investigación (España) and the Czech Science Foundation.

He is a member of different societies, such as the American Chemical Society, American Society for Mass Spectrometry and Sociedad Española de Química Analítica. He is also a member of the Spanish Network for Sample Preparation and the EuChemS-DAC Sample Preparation Task Force and Network.

Preface to "Five Years of Separations"

This Special Issue is dedicated to the celebration of five years of the Separations Journal and aims to collect original research papers from the frontiers of separation research, as well as review articles from prominent scholars, highlighting the state-of-the-art of separation science and technology. Researchers and technologists, whose work focuses on separations and related applications, were invited to contribute with papers disseminating their excellent research findings. The issue shows the multidisciplinary approach of separation techniques, and it is a forum to share innovative ideas in the field.

One review and nine original research articles are included. The Guest Editors wish to thank all authors for their fine contributions.

Separations **has already received its first Impact Factor, which is 1.9, proving that it is one of the most promising journals in the field of chromatographic techniques and separation science.**

Victoria Samanidou, Rafael Lucena
Editors

Review

Applications of Gas Chromatography for the Analysis of Tricyclic Antidepressants in Biological Matrices

Natalia Manousi and Victoria F. Samanidou *

Laboratory of Analytical Chemistry, Department of Chemistry, Aristotle University of Thessaloniki, 54124 Thessaloniki, Greece; nmanousi@chem.auth.gr
* Correspondence: samanidou@chem.auth.gr; Tel.: +30-2310-997-698

Received: 30 January 2019; Accepted: 24 April 2019; Published: 29 April 2019

Abstract: Tricyclic antidepressant drugs (TCAs) are a main category of antidepressants, which are until today widely used for the treatment of psychological disorders due to their low cost and their high efficiency. Therefore, there is a great demand for method development for the determination of TCAs in biofluids, especially for therapeutic drug monitoring. Gas chromatography (GC) was the first chromatographic technique implemented for this purpose. With the recent development in the field of sample preparation, many novel GC applications have been developed. Herein, we aim to report the recent application of GC for the determination of tricyclic antidepressants in biofluids. Emphasis is given to novel extraction techniques and novel materials used for sample preparation.

Keywords: gas chromatography; tricyclic antidepressants (TCAs); sample treatment; biological fluids

---

1. Introduction

Tricyclic antidepressant drugs (TCAs) are widely used for the treatment of psychiatric disorders such as depression [1]. They were firstly introduced in the 1950s with the discovery of imipramine by Roland Kuhn [2]. Due to their low cost and their high efficiency, they are widely prescribed until today for the treatment of major depression disorder, despite the introduction of newer antidepressants [1]. The chemical structures of TCAs are shown in Figure 1.

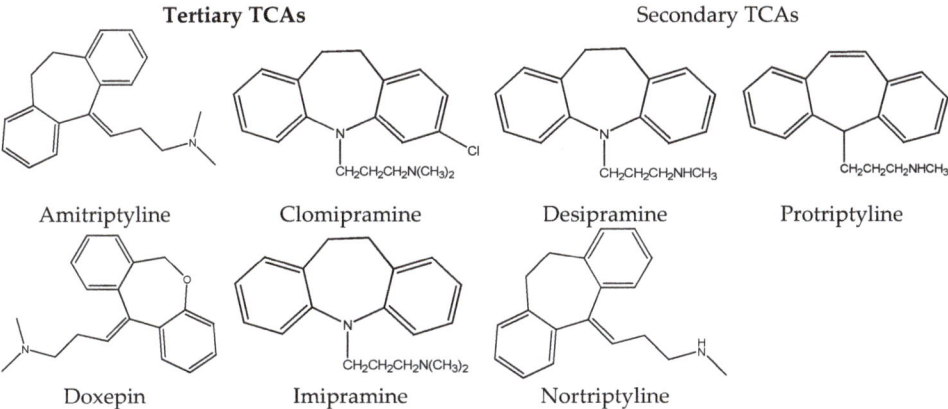

Figure 1. Chemical Structure of some tricyclic antidepressants.

The name of tricyclic antidepressants is based on their chemical structure, which contains three rings of atoms. Tricyclic antidepressants can be categorized as tertiary amine and as secondary amine TCAs.

Tertiary amine TCAs include amitriptyline, imipramine, clomipramine, dothiepin and doxepin while secondary amine TCAs include desipramine, nortriptyline, and protriptyline. They consist of three fused hydrocarbon rings linked to an alkylamine chain. Structural isomers of tricyclic antidepressants include N– and O–heteroatoms in the rings, hydrocarbon chain length and double bond positions. Tricyclic antidepressants are able to produce therapeutic responses in patients with major depression and hence are primarily used for its treatment [3–5].

Demethylation, hydroxylation and/or oxidation are the major processes for metabolites formation. Thus, polar metabolites are formed in the liver and then excreted to kidney with about 5% of the drug remaining unchanged. Aromatic hydroxylation takes place and the site of this biotransformation on the drug varies for different compounds. Accordingly, glucuronide conjugation results in more lipophilic and water-soluble tricyclics with more efficient renal excretion. Finally, a significant fraction of hydroxy metabolites is removed with urine [4].

TCAs' dose usually ranges from 75 to 300 mg/day. Unconjugated form TCAs concentrations in human fluids are measured for therapeutic drug monitoring (TDM). Since antidepressants are highly protein-bound, the therapeutic concentrations of the "free" drug are quite low. The measurement of the concentration of the free drugs in blood is very important for the determination of the pharmacological activity of TCAs. Therefore, there is a high demand for analytical purpose that can achieve this goal [5]. It has been found that the maximal therapeutic efficacy achieved with notriptyline is when plasma levels are 50–175 ng/mL. For the other TCAs, the picture is less clear, however the expected plasma concentration ranges 70–300 ng/mL. Toxic therapeutic dose is beyond 450 ng/mL. In urine, the expected concentration of TCAs and their metabolites ranges 500–5000 ng/mL, depending on the compound [6,7].

High performance liquid chromatography tandem with ultraviolet detector (UV), diode array detector (DAD) and Mass Spectrometry (MS) is nowadays the most famous technique for the determination of TCAs in biological matrices [8–10]. Moreover, with the use of liquid chromatography tandem mass spectrometry (LC-MS/MS), sensitivity of TCAs determination has significantly improved [11,12]. Gas chromatography was popular for this purpose before 1975 and is also gaining more and more attention again recently [1]. Other techniques that have been applied for the determination of TCAs are capillary electrophoresis (CE) [13], voltammetry [14], fluorescence polarization immunoassay [15], amperometry [16], flow injection analysis [17], biosensors [18], mass spectrometry [19], corona discharge ion mobility spectrometry [20], electrospray ionization-ion mobility spectrometry [21], turbulent-flow liquid chromatography-MS [22].

Among the studied biofluids, human plasma, serum and whole blood are the most common together with urine [4]. Other matrices include oral fluid, human hair and more recently dry blood spots have been also examined [4,23].

Lately, a lot of progress has been made in the field of sample preparation. Before 2008, protein precipitation, dilution, liquid-liquid extraction (LLE) and solid-phase extraction (SPE) were the most common sample preparation techniques for the analysis of biological matrices [1]. Novel techniques such as solid phase microextraction (SPME) [24], liquid phase microextraction (LPME) [4], extraction with QuEChERS [25], magnetic solid phase extraction (MSPE) [26], etc. have been recently applied to sample preparation of biological fluids. Moreover, novel materials such as metal–organic frameworks [26], functionalized Fe_3O_4 nanoparticles [27], paramagnetic core–shell functionalized nanoparticles [28], etc. have been also tested to replace conventional sorbents.

Only a limited number of review articles can be found in the literature concerning tricyclic antidepressants determination in biological matrices with gas chromatography. In 1980, Scoggins et al. wrote a review about the measurement of tricyclic antidepressants between 1967 and 1980 [29]. Gupta et al. wrote a review about the determination of tricyclic antidepressant drugs by gas chromatography with the use of a capillary column in 1983 [30]. The same year, Van Brunt published a review regarding the application of new technology for the measurement of tricyclic antidepressants using capillary gas chromatography with a fused silica DB5 column and nitrogen phosphorus detection [31].

In 1985, Norman published a review regarding the chromatographic techniques that have been implemented for the determination of TCAs in human plasma and human serum by chromatographic techniques [32]. Smyth discussed the applications of liquid chromatography–electrospray ionization mass spectrometry (LC–ESI-MS) to the detection and determination of TCAs in biological fluids and other matrices and made a comparison with gas–liquid chromatography–mass spectrometry (GLC–MS), when it was possible [33]. Since 2008, a few reviews have been published for the determination of TCAs with HPLC, however limited attention has been given to gas chromatography applications [1,4,34].

2. Early Use of Gas Chromatography

The first reported method for the determination of a tricyclic antidepressant drug was published in 1968. Stephen Curry developed a gas–liquid chromatography method for the determination of chlorpromazine and some of its metabolites in human plasma. For the separation, a packed column with 3% OV 17 phase (by Ohio Valley Specialty Company), which is a mid-polar phase containing 50% diphenyl and 50% dimethylpolysiloxane was used. Extraction with *n*-heptane was used for sample preparation and detection was achieved with an electron capture detector [35]. In 1975, Gifford et al. used a specific nitrogen detector for the determination of TCA drugs in plasma. Due to the sensitive alkali flame ionization detector no derivatization step was needed [36]. The same year, Aksel Jorgensen used a glass column filled with Chromosorb CQ (100–120 mesh) coated with 1% OV 17 for the determination of amitriptyline and nortriptyline in human Serum. Extraction was performed with *n*-hexane and limits of detection were 5 ng/mL for amitriptyline and 10–15 ng/mL for nortriptyline [37]. In 1976, Vasiliades et al. developed a gas liquid chromatographic determination of therapeutic and toxic levels of amitriptyline in human serum and limit of detection for amitriptyline was reduced to 1 ng/mL. For this purpose, the analytes were extracted from an alkaline solution into *n*-heptane containing 4% isobutanol and back-extracted into 0.1 M hydrochloric acid [38]. The same year, Bailey et al. developed a GC method for the determination of therapeutic concentrations of imipramine and desipramine in plasma. Imipramine was measured as the unchanged base while desipramine was measured as its *N*-trifluoroacetyl derivative. Prior to derivatization, the TCAs were extracted from an alkaline solution (pH 10.5 with Na_2CO_3) into hexane/isoamyl alcohol (98.5:1.5, *v/v*). That was the first reported application of derivatization of TCAs prior to their detection with a specific nitrogen detector [39]. Claeys et al. developed for the first time a gas chromatographic mass spectrometric method for the simultaneous measurement of imipramine and desipramine in plasma by selected ion recording with deuterated internal standards. The analytes were extracted with *n*-hexane and then derivatization with trifluoroacetylimidazole took place. Limits of quantification were reduced to nanogram level due to specificity provided by selected ion recording of the $[M + H]^+$ ions produced by chemical ionization using methane as reagent [40]. As can be easily observed, liquid–liquid extraction from biological samples, packed GC columns and selective nitrogen detectors were the most frequently used parameters for the analysis of biofluids for the determination of TCAs since early 1977. Helium was used in the most applications as carrier gas (mobile phase), followed by nitrogen [34–41].

At the end of 1977, mass spectrometry was also used for the simultaneous measurement of secondary and tertiary tricyclic antidepressants. For this purpose, tertiary amines such as amitriptyline, doxepin, and imipramine were analyzed underivatized, while secondary amines such as nortriptyline, desmethyldoxepin, desipramine, and protriptyline were analyzed after derivatization with trifluoroacetic anhydride. The analytes were extracted into a mixture of isopropanol/hexane (2:98, *v/v*) from the alkaline solution. Methane was used as carrier gas [42]. The same year, Garland developed a GC method for determination of amitriptyline and nortriptyline in human plasma. For the LLE procedure, *n*-hexane was chosen, while isobutane was chosen as carrier gas and reagent gas for chemical ionization [43]. In 1979, the same author developed a method for the determination of amitriptyline and its metabolites 10-hydroxyamitriptyline, 10-hydroxynortriptyline and nortriptyline in human plasma using stable isotope dilution and gas chromatography-chemical ionization mass spectrometry (GC-CIMS), using deuterated analogs as internal standards [44]. In 1979, Dhar et al. developed a gas–liquid chromatographic method for the determination of amitriptyline and nortriptyline levels in plasma using nitrogen-sensitive detectors after derivatization with trifluoroacetic anhydride [45]. In 1981, a nitrogen-phosphorous detector (NPD) was employed for the determination of imipramine, desipramine, doxepin, amitriptyline and nortriptyline. Limits of detection were reduced to 0.5–0.75 ng/mL. The analytes were extracted from alkaline solution into n-hexane–isoamyl alcohol mixture (98:2, v/v) [46]. The same year, Narasimhachari et al. developed a quantitative mapping of metabolites of imipramine and desipramine in plasma samples by gas chromatography–mass spectrometry with selected ion-monitoring (SIM) using deuterated analogues as internal standards. LLE was chosen for sample preparation in combination with derivatization with N-methyl-bis-trifluoroacetamide [47]. In 1982, Hals et al. developed a sensitive gas chromatographic assay for amitriptyline and nortriptyline in plasma and in 1983 Jones et al. developed a GC method for the quantification of amitriptyline, nortriptyline, and 10-hydroxy metabolite isomers in plasma [48,49]. In 1983, Gupta et al. used for the first time a DB-1 (30 m × 0.25 mm i.d.) capillary column for the determination of TCAs in plasma by GC coupled with a nitrogen selective detector. For the sample preparation, the samples were washed with pentane at acidic pH and extracted with pentane at alkaline pH [30]. In 1984, Ishida et al. developed a GC-MS method for the determination of amitriptyline and its major metabolites (nortriptyline, 10-hydroxyamitriptyline and 10-hydroxynortriptyline) in human serum using electron impact ionization. LLE was chosen for sample preparation and the analytes were extracted from an alkaline solution into n-hexane. Helium was chosen as carrier gas [50].

In 1990, surface ionization detector (SID) was introduced by Hattori et al. for the detection of tricyclic antidepressants in body fluids. A capillary SPB-1 GC column (30 m × 0.32 mm I.D., 0.25 µm) was used for the separation of the analytes and solid phase extraction was firstly proposed as a clean-up and preconcentration step for the extraction of biofluids. For this purpose, Sep-Pak C_{18} cartridges were pretreated with chloroform-2-propanol (9:1), acetonitrile and distilled water. Subsequently, the sample was loaded, and the cartridges were washed with water and finally the analytes were eluted withchloroform-2-propanol (9:1) [51].

In 1996, Ulrich et al. developed a gas chromatographic method for the simultaneous quantification of amitriptyline, nortriptyline and four hydroxy metabolites in human serum or plasma. The method was based on a three-step LLE. For the separation a HP-5 (25 m × 0.2 mm i.d., 0.33 µm) was employed [52].

In 1997, Pommier et al. used a capillary column for the simultaneous determination of imipramine and its metabolite desipramine in human plasma by gas chromatography coupled with mass-selective detection. The column was a fused-silica column coated with 5% phenyl methyl silicone (12 m × 0.2 mm i.d., 0.33 µm). The analytes were extracted at basic pH into n-heptane–isoamyl alcohol (99:1, v/v) [53]. The same year, Lee et al. developed a method for the detection of tricyclic antidepressants in whole blood by headspace solid-phase microextraction and capillary gas chromatography. For this purpose, the samples were heated at 100 °C in a septum-capped vial in the presence of distilled water and sodium hydroxide. A polydimethylsiloxane-coated SPME fiber was immersed to headspace of the vial to adsorb the analytes. For the detection, a flame-ionization detection (FID) was used. Recoveries were 5.3–12.9% [54].

In 1998, de la Torre et al. developed a capillary GC-NPD method for the quantitative determination of tricyclic antidepressants and their metabolites in human plasma by SPE Bond-Elut TCA). With this procedure, recoveries were higher than 87% [55]. The same year, Way et al. developed an isotope Dilution GC-MS method for the determination of TCA drugs in plasma. For the derivatization of secondary amine drugs carbethoxyhexafluorobutyryl chloride was examined in order to replace trifluoroacetyl and heptafluorobutyryl derivatives, which are relatively unstable and cause rapid deterioration of capillary GC columns. The obtained derivatives were stable and therefore this reagent can be utilized for TCAs derivatization [56].

As it can be observed, a lot of progress was made in the field of tricyclic antidepressants determination during 1968–2000. Various GC columns (either packed columns or capillary columns) have been used for the separation of TCAs. Moreover, different carrier gases (mobile phase) have been used. Helium and nitrogen were the most frequently chosen, while other gases such as methane and isobutane have been also used. As for the sample preparation, LLE extraction was by far the most famous method for the sample preparation of biofluids before 2000. Organic solvents such as n-heptane, *n*-hexane and mixtures such as *n*-heptane–isoamyl alcohol (99:1, *v/v*), *n*-hexane–isoamyl alcohol (98:2, *v/v*), etc. were examined. SPE applications with Sep-Pak C_{18} and Bond-Elut TCA cartridges and SPME applications are also reported in the literature. Various chemical reactions resulting in different trifluoroacetyl, heptafluorobutyryl or carbethoxyhexafluorobutyryl derivatives were also tested for sensitivity enhancement. Finally, regarding the detector system, NPD detectors and MS detectors are the most widely used detection systems for TCAs in biofluids. Other systems, including ECD, FID and nitrogen selective detectors, have also been employed.

Although in early 1970s many GC methods were reported in the literature, there was a lack of HPLC methods. In 1975, separation of TCAs with liquid chromatography was reported, however it was not until 1976 that the practicability of measuring clinical samples by HPLC was achieved [32,57]. However, between 1976 and 1985, liquid chromatography was increasingly applied to tricyclic antidepressants determination in biological matrices.

3. Recent Advances in the Use of Gas Chromatography

Due to the high increase in HPLC applications, there was a decreasing rate of application of gas chromatography between 1985 and 2000. However, GC is until today widely used for the determination of tricyclic antidepressants in biofluids. A lot of progress has been done especially in the field of sample preparation and many research papers are published in the literature after 2000.

3.1. Gas Chromatography-Mass Spectrometry Methods

In 2004, Paterson et al. developed a screening and semi-quantitative analysis of post mortem blood for basic drugs using GC-MS to evaluate whether 14 drugs (amitriptyline, citalopram, clozapine, cocaine, cyclizine, diazepam, dihydrocodeine, dothiepin, methadone, mirtazapine, procyclidine, sertraline, tramadol, and venlafaxine) were present in sub-therapeutic, therapeutic or greater than therapeutic concentration. For this purpose, liquid–liquid extraction was used for sample preparation. Blood samples were treated with ammonia for pH adjusting to 10. Subsequently, the analytes were extracted into diethylether and back extracted into 0.1 M HCl. For the separation a DB-5 (30 m × 0.25 mm, 0.25 μm) was used and helium was delivered at a flow rate of 1 mL/min. Under optimum conditions, LOQ for amitriptyline was 0.05 ng/mL. Additionally, trimipramine, desipramine and clomipramine could be detected, but they were not semi-quantified [58].

In 2006, Crifasi et al. examined the usefulness of twister bar extraction in combination with thermal desorption for basic drug screening of forensic samples by GC-MS. Research was also made for the investigation of the necessary conditions for basic drug isolation with stir bar sorptive extraction (SBSE). For this purpose, drugs of different categories, including TCAs were present in this study. Desorption was performed in the TDU unit with a helium flow of 50.0 mL/min a split ratio of 20:1. It was concluded that these kind of desorptive methods are as efficient as conventional LLE and SPE methods, while use of extraction solvents and complicated steps is avoided [59].

In 2007, Sarafraz-Yazdi et al. developed a GC-MS method for the determination amitriptyline and nortriptyline by directly suspended droplet microextraction prior to GC analysis of urine samples. For this purpose, basified samples were agitated with a stirring bar in order to create a mild vortex at the center of the vial and 10 μL of toluene was placed at the bottom of the vortex. Extraction was achieved in 20 min and the organic droplet was withdrawn with a syringe and analyzed. Recovery was 76.08% for amitriptyline and 82.62% for nortriptyline, while LOQs were 132–165 ng/mL for amitriptyline and 0.05 μg/mL for nortriptyline. A CP-Sil 24 CB (30 m × 0.32 mm, 0.25 μm) was used for separation and helium was delivered at a flow rate of 1.11 mL/min [60].

In 2008, Rana et al. developed a GC-MS method for the simultaneous determination of amitriptyline, nortriptyline, imipramine, desipramine, doxepin, desmethyldoxepin, and maprotiline in human urine after enzymatic hydrolysis with β-glucuronidase from *Escherichia coli*. Hydrolysis was performed to assist in the extraction procedure of tertiary TCAs, which are extensively conjugated in urine. Therefore, β-glucuronidase K12 from *Escherichia coli* were mixed with phosphate buffer and a portion of the mixture was added to urine. Incubation took place at 52 °C for 1 h. After cooling, the samples were transferred in tubes with a salt mixture (sodium chloride:sodium carbonate:sodium bicarbonate, 6:1:1 *w/w/w*) and the extraction solvent mixture (dichloromethane, dichloroethane, heptane and isopropyl alcohol (5:5:10:1 *v/v/v/v*). Subsequently, derivatization of the TCAs took place with MSTFA/ammonium iodide/ethanethiol reagent. For the GC analysis a CP-Sil 5 CB (10 m × 0.15 mm, 0.12 μm) column was chosen. The mobile phase was hydrogen and it was delivered at a flow rate of 1 mL/min. Two different oven temperature programs were used: one for doxepin and desmethyldoxepin and the other for the other analytes. LOQs were 5–100 ng/mL, while recoveries from amitriptyline, imipramine and doxepin were significantly increased after hydrolysis [61].

In 2008, Lee et al. developed a GC-MS method for the determination of four tricyclic antidepressants (amitriptyline, amoxapine, imipramine, and trimipramine) in human plasma using pipette tip solid-phase extraction with MonoTip C_{18} tips. For the sample pretreatment, human plasma containing protriptyline as internal standard was basified and centrifugated. For the SPE procedure, the sorbent was preconditioned twice with methanol and water using a manual micropipettor and the supernatant was extracted to by 20 repeated aspirating/dispensing. Elution was achieved with methanol by five repeated aspirating/dispensing cycles and the eluate was directly injected into a GC-MS system. A DB-5MS fused silica capillary column (30 m × 0.32 mm id, 0.25 μm) was used for the separation. Helium was used as a carrier gas at a flow rate of 2.0 mL/min. Recovery ranged 80.2–92.1% and LOQs were 0.2–5 ng/mL [62].

Dispersive liquid–liquid microextraction was successfully applied in the determination of TCAs in human urine by GC-MS after in situ derivatization. Urine samples were primarily treated with acetonitrile and their pH value was adjusted with sodium carbonate. For the DLLME procedure, methanol (disperser solvent), carbon tetrachloride (extraction solvent), and acetic anhydride (derivatization reagent) were injected rapidly into a human urine sample. The resulted sedimented phase that contained the derivatives of the TCAs was analyzed by GC-MS. A DB-5MS capillary column (30 m × 0.25 mm i.d., 0.5 μm) was used for separation. Helium delivered as the carrier gas at a flow rate of 1.0 mL/min. The average recoveries of TCAs were 88.2–104.3% and LOQs were 2–5 ng/mL. Compared to SPME or LPME method, DLLME method is rapid, simple, and inexpensive [63].

In 2011, Rani et al. developed a GC-MS method and an LC-UV method for the quantification of tricyclic and nontricyclic antidepressants in plasma and urine samples after microextraction in packed syringe. The studied TCAs were amitriptyline, imipramine and clomipramine. Plasma and urine samples were centrifuged and then aliquots of 50 μL were loaded into a Barrel Insert Needle Assembly (BIN) containing 4 mg of C_{18} sorbent conditioned with methanol and water. Subsequently, the sorbent was washed with water and the analytes were eluted with methanol. For the GC analysis, a Rtx-1MS (30 m × 0.25 mm id, 0.25 μm) was used. Helium was used as mobile phase at a flow rate of 1 mL/min. LOQs were 0.330–0.608 ng/mL and recoveries 77–99%. The developed method was compared with the developed LC-UV method and gave comparable accuracy and precision results. However, GC-MS has higher sensitivity, selectivity and capability of direct injection of samples into the mass spectrometer [64].

Papoutsis et al. developed a GC-MS method for the determination of different drugs and some of their metabolites (amitriptyline, citalopram, clomipramine, fluoxetine, fluvoxamine, maprotiline, desmethyl-maprotiline, mirtazapine, desmethyl-mirtazapine, nortriptyline, paroxetine, sertraline, desmethyl-sertraline, venlafaxine and desmethyl-venlafaxine) in whole blood. Bond Elut LRC Certify cartridges were used for the solid phase extraction of the analytes. An HP-5MS capillary column (30 m × 0.25 mm i.d., 0.25 μm) was used for the separation. Helium at a flow rate of 1 mL/min was used as mobile phase. For the sample preparation, blood samples were treated with phosphate buffer and centrifugation took place. The cartridges were conditioned with methanol and phosphate, the samples were loaded, and the sorbent was washed with water, acetic acid 1.0 M and methanol. Drying of the

SPE took place under vacuum and finally the analytes were eluted with a mixture of ethyl acetate: isopropanol: ammonium hydroxide (85:15:3, $v/v/v$). Evaporation and reconstitution were followed by derivatization with heptafluorobutyric anhydride. Recoveries ranged from 79.2% to 102.6% and LOQs were 1–5 ng/mL [65].

In 2013, Farag et al. developed a GC-MS method for the amitriptyline and imipramine in urine using clomipramine as internal standard. Liquid–liquid extraction was used for sample preparation and analytes were extracted from the alkaline pH into n-hexane–ethyl acetate (9:1, v/v) and back-extracted into acidic aqueous solution. Subsequently, derivatization with BSTFA-1% TMCS was performed 60 °C for 30 min. For the separation, a DB-5MS column (30 m × 0.25 mm i.d, 0.5 µm) was used and helium was delivered as carrier gas at flow rate of 1 mL/min. Recovery values were higher than 89.7% and LOQs were 100 µg/mL for both analytes and the IS [66].

A hollow-fiber liquid–phase microextraction (HF-LPME) was used for the determination of amitriptyline, nortriptyline, imipramine, desipramine, clomipramine, desmethylclomipramine, fluoxetine, and norfluoxetine in whole blood by GC-MS. Optimum conditions for sample preparation were as follows: a disposable 8-cm polypropylene porous hollow fiber, 4.0 mL of sample solution, dodecane as organic phase, and 0.1 M formic acid as acceptor phase for extraction. The system was stirred, and the acceptor phase was evaporated and reconstituted in methanol. An HP-5MS column (30 m × 0.25 mm i.d., 0.25 µm) was chosen for separation. Helium was used as carrier gas at a flow rate of 0.8 mL/min while splitless injection mode was chosen. LOQs were 20 ng/mL and recoveries ranged 36–89% [67].

In 2014, Banitaba et al. used a new fiber coating based on electrochemically reduced graphene oxide for the cold-fiber headspace SPME of amitriptyline, imipramine and clomipramine in plasma. For the HS-SPME, solution pH of 13, NaCl content of 5% w/v, extraction time of 60 min at 70 °C was selected as the optimum salt content. The effect of the extraction time on the HS-SPME of A CP-Sil 8 CB (25 m × 0.32 mm id, 0.25 µm) was chosen in combination with nitrogen as carrier gas delivered at a constant pressure of 17 psi. LOQs were 1.0–1.7 ng/mL and recoveries varied from 73% to 96% [68].

In 2018, Mohebbi et al. developed a dispersive solid phase extraction combined with deep eutectic solvent-based air-assisted liquid–liquid microextraction for the extraction of amitriptyline, nortriptyline, imipramine, desipramine and clomipramine from plasma and urine by GC-MS. Therefore, a sorbent (C_{18}) was dispersed by vortex into an alkaline sample solution, the material was isolated by centrifugation and the analytes were eluted with 150 µL of a deep eutectic solvent, prepared from choline chloride and 4-chlorophenol. Subsequently, the eluent was mixed with ammoniacal buffer and rapidly injected with the assist of a syringe into alkaline deionized water. Five aspiration/dispersion cycles were performed and after centrifugation 1 µL of the sedimented deep eutectic solvent phase was analyzed by GC-MS. For the separation, an HP-5MS column (60 m × 0.25 mm i.d., 0.25 µm) was used. Extraction recoveries were 62–74% for urine and 64–72% for plasma and LOQs were 27–49 ng/L for urine and 108–191 ng/L for plasma [69].

In 2018, Japtag et al. developed a hollow-fiber drop-to-drop solvent microextraction technique for the extraction of nortriptyline from urine and blood by GC-MS. Therefore, a glass syringe was filled with the organic acceptor phase which was inserted into the hollow fiber. Thus, the extraction of the drug was performed with 0.6 µL of toluene as a solvent for 12 min extraction time by immersing the fiber in the aqueous sample solution (pH 9.8). For the separation, a DB-5 capillary column (30 m × 0.25 mm × 1 µm) with the assist of helium (flow: 1 mL/min) LOQ value was 234 ng/mL and recovery ranged from 97.33% to 103.66% [70].

3.2. Gas Chromatography with Various Detector Systems

In 2002, Martinez et al. developed a GC-NPD method for the simultaneous determination of imipramine and desipramine together with viloxazine, venlafaxine, sertraline, and amoxapine in whole blood. For this purpose, two different SPE cartridges (Chem Elut SPE and Bond Elut Certify) were tested. Chem Elut cartridges are based on the principle of liquid–solid absorption extraction that is related to conventional LLE. On the other hand, Bond Elut Certify columns take advantage of mixed SPE, reversed-phase and cation-exchange sorbent. With Chem Elut cartridges, recoveries were between 28% and 74% and limits of detection (LOD) ranged from 39 to 153 ng/mL, while, with Bond Elut Certify columns, recoveries were between 64% and 86% and LOQs ranged from 70 to 222 ng/mL. Therefore, higher recoveries, lower LODs, cleaner extracts, and better sensitivity and precision, together with less solvent consumption were obtained with Bond Elut Certify. For the separation, an HP-1 (25 m × 0.20 mm i.d., 0.11 μm) was employed. Carrier gas was helium and it was delivered at a constant pressure of 195 kPa. Instrumental analysis parameters, recovery values and LODs of the developed methods are presented at Table 1 [71].

One year later, the same working group compared the two SPE columns for the determination of amitriptyline, nortriptyline, trimipramine, and clomipramine together with fluoxetine maprotiline and trazodone in whole blood by GC-NDP. In this case, recoveries of the analytes using Chem Elut columns were 30–50% and LOQs were between 44 and 485 ng/mL, while with Bond Elut Certify columns recoveries were 59–84% and LODs were between 8 and 67 ng/mL. Therefore, Bond Elut Certify are columns have the same benefits for the extraction of the examined analytes as in the previous study [72].

Yazdi et al. developed a GC-FID method for the determination of amitriptyline and nortriptyline from environmental solutions after dispersive liquid-liquid microextraction. For the sample preparation, carbon tetrachloride (extraction solvent) and methanol (disperser solvent) were injected rapidly into the aqueous sample and a cloudy solution was formed. After centrifugation, the extraction solvent was sedimented on the bottom of the conical test tube and it was removed for GC analysis on a CP-Sil 24 CB (30 m × 0.32 mm, 0.25 μm) capillary column. The performance of the proposed technique was also evaluated for determination of TCAs in blood plasma. However, it was not found to be compatible for extraction from biological samples, due to the interaction of matrix components, which did not allow the carbon tetrachloride to produce a sedimented phase for injection to GC [73].

Electromembrane SPME was also used for the extraction of amitriptyline and doxepin from water, human plasma, and urine. The TCAs were extracted from 24 mL neutral sample solution across organic liquid membrane and into aqueous acceptor phase with the application of 120 V electrical potential for 20 min and the analytes were adsorbed on a carbonaceous cathode at a stirring rate of 1250 rpm. Direct desorption into the GC port took place at 280 °C for 2 min. Extraction efficiencies were 3.1–11.5% and quantification limits were less than 5 ng/mL. An HP-5 (30 m × 0.32 mm i.d., 0.25 μm) was used with helium at constant flow rate of 0.6 mL/min [74].

In 2011, Yazdi et al. developed a GC-FID method for the separation and determination of amitriptyline and nortriptyline in biological samples after single-drop microextraction [75]. Single-drop microextraction (SDME) is a variation of LPME, that was firstly introduced in 1996 by Jeannot and Cantwell [76]. For the SDME procedure, the sample solution was kept alkaline (pH 12), then a microdroplet of isooctane was exposed in the stirred solution for 15 min. When the extraction was finished, the droplet was retracted back into the syringe and injected directly into the GC system. A CP-Sil 24 CB capillary column (30 m × 0.32 mm id, 0.25 μm) was used for the separation in combination with helium as carrier gas delivered at a flow rate of 1.11 mL/min. Recoveries of amitriptyline were 78.65% in plasma and 92.07% in urine, while for nortriptyline were 66.54% in plasma and 97.39% in urine. LOQs were 0.033–0.066 μg/mL [75].

Table 1. Instrumental analysis parameters, sample preparation techniques, recovery values and LODs of the developed GC methods during 2000–2019.

TCAs	Matrix	Sample Preparation	GC Parameters	Detector	LOQs (ng/mL)	Recoveries	Ref.
Imipramine, desipramine	Plasma	MSPE	Column: CP-Sil 8 CB column (30 m × 0.25 mm i.d., 0.25 μm) Carrier gas: He (pressure of 10 psi) Injection mode: splitless Injector: 270 °C and Detector: 270 °C Oven: 80 °C (1 min), 30 °C/min to 280 °C (5 min).	FID	10–4000	94%	[28]
Amitriptyline	Blood	LLE	Column: DB-5 (30 m × 0.25 mm, 0.25 μm) Carrier gas: He (flow rate of 1 mL/min) Injection mode: Splitless Injector temperature: 280 °C and Detector: 300 °C Oven: 50 °C (2 min), 30 °C/min to 180 °C, 5 °C/min to 280 °C (19 min)	MS	NA	NA	[58]
Amitriptyline, nortriptyline	Urine	directly suspended droplet microextraction	Column: CP-Sil 24CB (30 m × 0.32 mm, 0.25 μm) Carrier gas: He was delivered at a flow rate of 1.11 mL/min. Injection mode: split ratio of 46.04/1 Injector: 280 °C and Detector: 280 °C Oven: 100 °C (1 min), 50 °C/min to 240 °C, 2 °C/min to 260 °C	MS	132–165	76.1–82.6%	[60]
Amitriptyline, nortriptyline, imipramine, desipramine	Urine	LLE after enzymic hydrolysis	Column: CP-SIL 5CB (10 m × 0.15 mm, 0.12 μm) Injection mode: Splitless Carrier gas: H₂ delivered at a flow rate of 1 mL/min. Injector: 250 °C Oven: two different acquisition programs	MS	5–100	NA	[61]
amitriptyline, amoxapine, imipramine, trimipramine	Plasma	pipette tip SPE	Column: DB-5MS fused silica capillary column (30 m × 0.32 mm id, 0.25 μm) Carrier gas: He (flow rate of 2.0 mL/min). Injection mode: splitless Injector: 250 °C and Detector: 280 °C Oven: 100 °C (1 min), 20 °C/min to 300 °C.	MS	0.2–5	80.2–92.1%	[62]
Imipramine, desipramine, amitriptyline, nortriptyline, clomipramine	Urine	DLLME	Column: DB-5MS (30 m × 0.25 mm i.d., 0.5 μm) Carrier gas: He (flow rate of 1.0 mL/min) Oven temperature program: 60 °C (3 min), 15 °C/min to 300 °C (4 min).	MS	2–5	88.2–104.3%	[63]

Table 1. *Cont.*

TCAs	Matrix	Sample Preparation	GC Parameters	Detector	LOQs (ng/mL)	Recoveries	Ref.
Amitriptyline, imipramine, clomipramine	Urine, plasma	Microextraction by packed syringe	Column: Rtx-1MS (30 m × 0.25 mm id, 0.25 μm) Carrier gas: He (flow rate of 1 mL/min) Injection mode: split (10:1) Injector: 270 °C and Detector: 300 °C. Oven: 100 °C (1 min), 10 °C/min to 200 °C, 15°C/min to 260 °C, 30 °C/min to 300 °C.	MS	0.330–0.608	77–99%	[64]
Amitriptyline, clomipramine, nortriptyline	Whole blood	SPE	Column: HP-5MS capillary column (30 m × 0.25 mm i.d., 0.25 μm) Carrier gas: He (flow rate of 1 mL/min) Injector: 280 °C and Detector: 300 °C. Oven: 100 °C (1 min), 40 °C/min to 300 °C (4 min)	MS	1–5	79.2–102.6%	[65]
Amitriptyline, imipramine	Urine	LLE	Column: DB-5MS column (30 m × 0.25 mm i.d, 0.5 μm Carrier gas: Helium was delivered as carrier gas at flow rate of 1 mL/min Oven: 80 °C held for 3 min then raised to 300 °C at a rate of 30 °C and held for 4 min	MS	100000	89.7	[66]
Amitriptyline, nortriptyline, imipramine, desipramine, clomipramine	Whole blood	HF-LPME	Column: HP-5MS column (30 m × 0.25 mm i.d., 0.25 μm) Carrier gas: He (flow rate of 0.8 mL/min) Injection mode: splitless Injector: 280 °C Oven: 125 °C (1 min), 50 °C/min to 190 °C, 5 °C/min to 225 °C (3 min), 50 °C/min to 230 °C (1 min).	MS	20	36–89%.	[67]
Amitriptyline, imipramine clomipramine	Plasma	HS-SPME	Column: CP-Sil 8 CB (25 m × 0.32 mm id, 0.25 μm) Carrier gas: N_2 (pressure of 17 psi) Injector port: 280 °C and Detector: 300 °C Oven: 80 °C (3 min), 30 °C/min to 180 °C, 5 °C/min to 210 °C, 30 °C/min to 300 °C (1 min).	FID	1.0–1.7	73–96%	[68]
Amitriptyline, nortriptyline, imipramine, desipramine, clomipramine	Plasma, urine	d-SPE and deep eutectic solvent-based air-assisted LPME	Column: HP-5MS column (60 m × 0.25 mm i.d., 0.25 μm). Carrier gas: He (flow rate of 1 mL/min). Detector: 260 °C Column: 100 °C (1 min), 50 °C/min to 190 °C, 5 °C/min to 225 °C (3 min), 20 °C/min to 300 °C (5 min).	MS	0.027–0.191	62–74%	[69]
Notriptyiline	Urine, blood	hollow-fiber drop-to-drop solvent microextraction	Column: DB-5 (30 m × 0.25 mm i.d., 1 μm) Carrier gas: of helium (flow: 1 mL/min) Injection mode: splitless Injector: 250 °C and Detector: 300 °C Oven: 80 °C (3min), 20 °C/min to 250 °C (3 min).	MS	234	97.33–103.66%	[70]

Table 1. Cont.

TCAs	Matrix	Sample Preparation	GC Parameters	Detector	LOQs (ng/mL)	Recoveries	Ref.
Imipramine, desipramine	Whole Blood	SPE	Column: HP-1 (25 m × 0.20 mm i.d., 0.11 µm) Carrier gas: He (constant pressure of 195 kPa) Injection mode: split (split ratio of 1:20) Injector: 280 °C and Detector: 300 °C Oven: 180 °C (1 min), 10 °C/min to 300 °C (3 min)	NPD	70–222	64–86%	[71]
Amitriptyline, nortriptyline, trimipramine, clomipramine	Whole Blood	SPE	Column: HP-1 (25 m × 0.20 mm i.d., 0.11 µm) Carrier gas: He (constant pressure of 195 kPa) Injection mode: split (split ratio of 1:20) Injector: 280 °C and Detector: 300 °C Oven: 180 °C (1 min), 10 °C/min to 300 °C (3 min).	NPD	44–485	59–84%	[72]
Amitriptyline	Plasma, urine	Electromembrane SPME	Column: HP-5 (30 m × 0.32 mm i.d., 0.25 µm) Carrier gas: He (constant flow rate of 0.6 mL/min Injector: 280 and Detector: 300 °C Oven: 160 °C (3 min), 30 °C/min to 280 °C (3 min)	FID	5	3.1–11.5%	[74]
Amitriptyline, nortriptyline	Plasma, urine	Single-drop microextraction	Column: CP-Sil 24 CB (30 m × 0.32 mm id, 0.25 µm Carrier gas: He (flow rate: 1.11 mL/min) Injection mode: split (1:46) Injector: 280 °C and Detector: 280 °C Oven: 100 °C (1 min), 20 °C/min to 240 °C, 2 °C/min to 260 °C.	FID	33–66	66.5–97.4%	[75]
Imipramine, clomipramine	Plasma, urine	EME	Column: CP-Sil 8 CB column (25 m × 0.32 mm i.d., 0.25 µm) Carrier gas: N$_2$ (constant pressure of 20 psi). Injection mode: splitless Injector: 270 °C and Detector: 300 °C. Oven: 140 °C, 32 °C/min to 280 °C (of 32 °C/min and held at 280 °C (1 min)	FID	2.3	90–95%	[77]
Desipramine	Plasma, urine	HF-LPME	Column: DB-35MS (30 m × 0.25 mm, 0.15 µm) Carrier gas: N$_2$ (head pressure of 0.12 MPa) Injection mode: splitless Injector: 260 °C and Detector: 310 °C Oven: 100 °C (1 min), 30 °C/min to 300 °C (9 min).	NPD	66	32%	[78]
Imipramine, clomipramine, desipramine	Urine	electro-assisted SPME	Column: CP-Sil 8 CB column (25 m × 0.32 mm i.d., 0.25 µm) Carrier gas: nitrogen as carrier gas, delivered with constant pressure of 16 psi Injector: 270 °C and Detector: 300 °C Oven: 100 °C, 30 °C/min to 300 °C (1 min)	FID	0.5–1.5	16–74.2%	[79]

Table 1. Cont.

TCAs	Matrix	Sample Preparation	GC Parameters	Detector	LOQs (ng/mL)	Recoveries	Ref.
Amitriptyline, trimipramine	Plasma, urine	EME and DLLME	Column: HP-5 (30 m × 0.32 mm i.d., 0.25 µm) Carrier gas: He (constant flow rate of 4 mL/min) Injection mode: split, ratio 1:5 Injector: 280 °C and Detector: 300 °C Oven: 185 °C (12 min), 30 °C/min to 280 °C (3 min).	FID	10 and 40	88.8–92.3%	[80]
Amitriptyline, imipramine	Plasma	air-agitated liquid–liquid microextraction	Column: BP-5 capillary column (30 m × 0.32 mm i.d., 0.25 µm). Injection mode: splitless Injector: 280 °C and FID: 280 °C Oven: 100 °C (3 min), 20 °C/min to 280 °C (2 min).	FID	15–20	68–73%	[81]

In 2012, Davarani et al. developed an electromembrane extraction combined with GC-FID for quantification of imipramine and clomipramine in human body fluid. Therefore, the TCAs were extracted from aqueous sample solutions, through a supported liquid membrane consisting of 2-nitrophenyl octyl ether impregnated on the walls of the hollow fiber. Optimum conditions were as follows applied voltage: 200 V, pH of donor solution: 2.0, pH of acceptor solution: 1.0, sample volume: 2.1 mL, acceptor volume: 7 µL, extraction time: 10 min, and stirring rate: 1400 rpm. A CP-Sil 8 CB column (25 m × 0.32 mm i.d., 0.25 µm) was used for the separation with nitrogen as carrier gas delivered at a constant pressure of 20 psi. LOQ was 2.3 ng/mL and recovery ranged 90–95% [77].

In 2012, Saraji et al. developed a HF-LPME method combined with in-syringe Derivatization with acetic anhydride for the extraction of desipramine from biological samples prior to GC-NPD analysis. For the HF-LPME procedure, n-dodecane was impregnated in the pores of the hollow fiber as the donor phase while methanol was placed inside the lumen of the fiber as the acceptor phase. Extraction of desipramine from the donor phase to the acceptor phase was achieved with stirring in 25 min at room temperature. Subsequently, the fiber was removed, the acceptor phase was withdrawn into the syringe and derivatization reagent was drawn into the syringe. Separation was achieved on a DB-35MS (30 m × 0.25 mm, 0.15 µm) stationary phase with nitrogen as carrier gas delivered at a head pressure of 0.12 MPa. LOQ were 66 ng/mL with and relative recovery was 32% [78].

In 2013, Davarani et al. used electro-assisted solid-phase microextraction for the sample preparation of urine samples for the quantification of imipramine, clomipramine and desipramine by GC-FID. Therefore, a platinum wire coated with poly(3,4-etylenedioxythiophen) that is able to adsorb both the ionic and molecular forms of the analytes was employed. For the sample preparation, sample solution was placed into the vial with 24% NaCl and the polymeric fiber (cathode) together with a platinum wire (anode), while the solution was stirred at 1000 rpm. The applied voltage was 2 V, the extraction temperature was 35 °C and extraction time was 20 min. When the extraction was finished, the fiber was inserted into the injector of GC for desorption. A CP-Sil 8 CB column (25 m × 0.32 mm i.d., 0.25 µm) column was chosen in combination with nitrogen as carrier gas, delivered with constant pressure of 16 psi. LOQs were 0.5–1.5 ng/mL and recoveries in urine ranged 16–74.2% [79].

In 2013, For the first time, combination of electromembrane extraction (EME) and dispersive liquid–liquid for the determination of amitriptyline, doxepine and trimipramine in human plasma and urine by GC-FID. Experimental design with response surface methodology was implemented for optimization of experimental parameters. As a result, extraction time of 14 min, applied voltage of 240 V, donor phase of 64 mM HCl and acceptor phase of 100 mM HCl were chosen for the EME process. For the DLLME process, methanol was chosen as disperser solvent and carbon tetrachloride was chosen as an extraction solvent. Centrifugation was necessary to separate the two phases of the resulting emulsion, prior to GC analysis. An HP-5 (30 m × 0.32 mm i.d., 0.25 µm) was used in combination with helium gas at constant flow rate of 4 mL min^{-1}. LOQs were 10 and 40 ng/mL in urine and plasma, respectively. The recoveries were 92.3–94.2% for urine and 88.8–92.6% for plasma [80].

In 2017, Asghari et al. applied for the extraction of amitriptyline and imipramine from human plasma and by air-agitated liquid–liquid microextraction. Therefore, the analytes were extracted by repeated aspiration and dissension of a solidifiable organic solvent, in the absence of an organic disperser solvent. Under optimum conditions, 14.0 µL of 1-dodecanol was used to extract the analytes from 10 mL of the sample (pH 12.0) after salt addition (7.52%, w/v) and 13 air-agitation cycles took place using a syringe. For the GC-FID analysis, helium was used as mobile phase at a flow rate of 4mL/min together with a BP-5 capillary column (30 m × 0.32 mm i.d., 0.25 µm). Low quantification limits 15–20 ng/mL and satisfactory recovery values (68–73%) were obtained [81].

In 2018, Ahmadi et al. synthesized Fe_3O_4 super paramagnetic core–shells anchored onto silica grafted with C_8/NH_2 nanoparticles and used them for the ultrasound-assisted magnetic solid phase extraction of imipramine and desipramine from plasma prior to GC-FID analysis. For the MSPE procedure, the material was washed and conditioned with water and MeOH and were dispersed by ultrasonic into the sample. Accordingly, a magnet was used to collect the sorbent and elution was achieved with methanol. A CP-Sil 8 CB column (30 m × 0.25mm i.d., 0.25 μm) was chosen and helium was delivered at a constant pressure of 10 psi. Recovery was above 94% and LOD values were 0.003–0.007 μg/mL [28].

4. Discussion

A lot of progress was made in this scientific field during 1980–2000. Both packed columns and capillary columns were used for this purpose. Moreover, various detection systems such as NPD, FID, SID, MS, were adopted. Liquid–liquid extraction was the most frequent sample preparation technique, while various solvents were tested. Derivatization with various reagents, resulting in different trifluoroacetyl, heptafluorobutyryl or carbethoxyhexafluorobutyryl, were also employed. Other conventional sample preparation techniques including protein precipitation and solid phase extraction have also been used.

However, due to the rapid development of high-performance liquid chromatography, the application of gas chromatography was significantly reduced after 2000. This can be attributed to the fact that HPLC, especially when coupled with mass spectrometry, can provide highly sensitive methods and extremely low limits of detection. Until today, HPLC is a well-established technique for the determination of various drugs (antidepressants, antipsychotics, antiepileptics, etc.).

The recent advances in the use of gas chromatography for the determination of TCAs in biofluids focus on the implementation of novel sample preparation techniques, such as LPME, SPME, EME, etc. Additionally, novel materials have been synthesized and tested for the extraction of tricyclic antidepressants from biofluids. These techniques follow the principles of "green chemistry" and have numerous advantages compared to conventional sample preparation methods.

The reported GC methods can be classified into the more sensitive GC-MS methods and the GC methods coupled with other detection systems (mostly ECD, NPD, and FID). The GC-MS methods combine the resolution ability of gas chromatography with the highly sensitive detection ability of mass spectrometry. Therefore, highly sensitive methods can be developed and low LODs can be obtained. Satisfactory LOQ values can be obtained both with GC-FID and with GC-NPD techniques. As shown in Table 1, by using a preconcentration method in combination with GC-MS, limits of quantification can be reduced to 0.2 ng/mL, while, with GC-FID and GC-NPD, they are more than 0.5 ng/mL and 44 ng/mL, respectively. This is of high importance especially for forensic toxicology applications in which highly sensitive and specific GC-MS (or LC-MS, LC-MS/MS) are required.

5. Conclusions

Gas chromatography is a well-established technique for the determination of TCAs in biological matrices. GC methods are considered rapid, simple, accurate, efficient and robust. In recent years, a lot of progress has been made in the field of sample preparation to replace the conventional techniques (protein precipitation, SPE, and LLE). For this purpose, various microextraction techniques such as LPME, SPME, MSPE, etc. have been implemented. These techniques are in compliance with the principles of "green chemistry". Moreover, novel materials have been synthesized and used to enhance preconcentration factors and to develop accurate methods with low limits of detection, which together with high recovery values can be achieved by GC, while sensitivity and selectivity can be higher than those of HPLC. However, a wide variety of materials such as metal–organic frameworks, graphene-oxide nanoparticles, etc. can be tested for such purposes. Additionally, recently developed microextraction techniques such as fabric phase microextraction and capsule phase microextraction can be also tested [82,83].

Funding: This research received no external funding.

Conflicts of Interest: The authors declare no conflict of interest.

References

1. Samanidou, V.; Nika, M.; Papadoyannis, I. HPLC as a Tool in Medicinal Chemistry for the Monitoring of Tricyclic Antidepressants in Biofluids. *Mini Rev. Med. Chem.* **2008**, *7*, 256–275. [CrossRef]
2. Brown, W.A.; Rosdolsky, M. The clinical discovery of imipramine. *Am. J. Psychiatry* **2015**, *172*, 426–429. [CrossRef] [PubMed]
3. Gillman, P.K. Tricyclic antidepressant pharmacology and therapeutic drug interactions updated. *Br. J. Pharmacol.* **2007**, *151*, 737–748. [CrossRef] [PubMed]
4. Uddin, M.; Samanidou, V.F.; Papadoyannis, I.N. Bio-sample preparation and analytical methods for the determination of tricyclic antidepressants. *Bioanalysis* **2011**, *3*, 97–118. [CrossRef] [PubMed]
5. Gupta, M.; Jain, A.; Verma, K.K. Determination of amoxapine and nortriptyline in blood plasma and serum by salt-assisted liquid–liquid microextraction and high-performance liquid chromatography. *J. Sep. Sci.* **2010**, *33*, 3774–3780. [CrossRef]
6. Melanson, S.E.F.; Tewandrowski, E.L.; Griggs, D.A.; Flood, J.G. Interpreting Tricyclic Antidepressant Measurements in Urine in an Emergency Department Setting: Comparison of Two Qualitative Point-of-Care Urine Tricyclic Antidepressant Drug Immunoassays with Quantitative Serum Chromatographic Analysis. *J. Anal. Toxicol.* **2007**, *31*, 270–275. [CrossRef]
7. Risch, S.C.; Huey, L.Y.; Janowsky, D.S. Plasma levels of tricyclic antidepressants and clinical efficacy: Review of the literature—part II. *J. Clin. Psychiatry* **1979**, *40*, 58–69. [PubMed]
8. Mohebbi, A.; Farajzadeh, M.A.; Yaripour, S.; Mogaddam, M.R.A. Determination of tricyclic antidepressants in human urine samples by the three-step sample pretreatment followed by HPLC-UV analysis: An efficient analytical method for further pharmacokinetic and forensic studies. *EXCLI J.* **2008**, *17*, 952–963. [CrossRef]
9. Uddin, M.N.; Samanidou, V.F.; Papadoyannis, I.N. Simultaneous Determination of 1,4-Benzodiazepines and Tricyclic Antidepressants in Saliva after Sequential SPE Elution by the Same HPLC Conditions. *J. Chin. Chem. Soc.* **2011**, *58*, 142–154. [CrossRef]
10. Coulter, C.; Taruc, M.; Tuyay, J.; Moore, C. Antidepressant Drugs in Oral Fluid Using Liquid Chromatography–Tandem Mass Spectrometry. *J. Anal. Toxicol.* **2010**, *34*, 65–72. [CrossRef]
11. Sempio, C.; Morini, L.; Vignali, C.; Groppi, A. Simple and sensitive screening and quantitative determination of 88 psychoactive drugs and their metabolites in blood through LC-MS/MS: Application on postmortem samples. *J. Chromatogr. B Anal. Technol. Biomed. Life Sci.* **2014**, *970*, 1–7. [CrossRef] [PubMed]
12. Fisichella, M.; Morini, L.; Sempio, C.; Groppi, A. Validation of a multi-analyte LC-MS/MS method for screening and quantification of 87 psychoactive drugs and their metabolites in hair. *Anal. Bioanal. Chem.* **2014**, *406*, 3497–3506. [CrossRef]
13. Acedoo-Valenzuela, M.; Mora-Diez, N.; Galeano-Diaz, D.; Silva-Rodriguez, A. Determination of Tricyclic Antidepressants in Human Breast Milk by Capillary Electrophoresis. *Anal. Sci.* **2010**, *26*, 26699–26702. [CrossRef]
14. Wang, J.; Golden, T.; Ozsoz, M.; Lu, Z. Sensitive and selective voltammetric measurements of tricyclic antidepressants using lipid-coated electrodes. *Bioelectrochem. Bioenerg.* **1990**, *23*, 217–226. [CrossRef]
15. Rao, M.L.; Staberock, U.; Baumann, P.; Hiemke, C.; Deister, A.; Cuendet, C.; Amey, M.; Hartter, S.; Kraemer, M. Monitoring tricyclic antidepressant concentrations in serum by fluorescence polarization immunoassay compared with gas chromatography and HPLC. *Clin. Chem.* **1994**, *40*, 929–939. [CrossRef] [PubMed]
16. Kataky, R.; Palmer, S.; Parker, D.; Spurling, D. Alkylated cyclodextrin-based potentiometric and amperometric electrodes applied to the measurement of tricyclic antidepressants. *Electroanalysis* **1997**, *9*, 1267–1272. [CrossRef]
17. Acedo-Valenzuela, M.I.; Galeano-Diaz, T.; Mora-Diez, N.; Silva-Rondriguez, A. Response surface methodology for the optimisation of flow-injection analysis with in situ solvent extraction and fluorimetric assay of tricyclic antidepressants. *Talanta* **2005**, *66*, 952–960. [CrossRef] [PubMed]
18. Knihnicki, P.; Wieczorek, M.; Moos, A.; Koscielniak, P.; Wietecha-Posłuszny, R.; Wozniakiewicz, M. Electrochemical sensor for determination of desipramine in biological material. *Sens. Actuators B* **2013**, *189*, 37–42. [CrossRef]
19. Santos, M.G.; Tavares, I.M.C.; Barbosa, A.F.; Bettini, J.; Figueiredo, E.C. Analysis of tricyclic antidepressants in human plasma using online-restricted access molecularly imprinted solid phase extraction followed by direct mass spectrometry identification/quantification. *Talanta* **2017**, *163*, 8–16. [CrossRef]

20. Aladaghlo, Z.; Fakhari, A.R.; Hasheminasab, K.S. Application of electromembrane extraction followed by corona discharge ion mobility spectrometry analysis as a fast and sensitive technique for determination of tricyclic antidepressants in urine samples. *Microchem. J.* **2016**, *129*, 41–48. [CrossRef]
21. Jafari, M.T.; Saraji, M.; Sherafatm, H. Electrospray ionization-ion mobility spectrometry as a detection system for three-phase hollow fiber microextraction technique and simultaneous determination of trimipramine and desipramine in urine and plasma samples. *Anal. Bioanal. Chem.* **2011**, *399*, 3555–3564. [CrossRef]
22. Breaud, A.R.; Harlan, R.; Di Bussolo, J.M.; McMillin, G.A.; Clark, W. A rapid and fully-automated method for the quantitation of tricyclic antidepressants in serum using turbulent-flow liquid chromatography–tandem mass spectrometry. *Clin. Chim. Acta* **2010**, *411*, 825–832. [CrossRef] [PubMed]
23. Berm, E.J.J.; Paardekooper, J.; Brummel-Mulder, E.; Hak, E.; Wilffert, B.; Maring, J.G. A simple dried blood spot method for therapeutic drug monitoring of the tricyclic antidepressants amitriptyline, nortriptyline, imipramine, clomipramine, and their active metabolites using LC-MS/MS. *Talanta* **2015**, *134*, 165–172. [CrossRef] [PubMed]
24. Alidoust, M.; Seidi, S.; Rouhollahi, A.; Shanehsaz, M. In-tube electrochemically controlled solid phase microextraction of amitriptyline, imipramine and chlorpromazine from human plasma by using an indole-thiophene copolymer nanocomposite. *Microchim. Acta* **2017**, *184*, 2473–2481. [CrossRef]
25. Alves, V.; Conceicao, C.; Goncalves, J.; Teixeira, H.M.; Camara, J.S. Improved Analytical Approach Based on QuECHERS/UHPLC-PDA for Quantification of Fluoxetine, Clomipramine and their Active Metabolites in Human Urine Samples. *J. Anal. Toxicol.* **2016**, *41*, 45–53. [CrossRef]
26. Safari, M.; Shahlaei, M.; Yamini, Y.; Shakorian, M.; Arkan, E. Magnetic framework composite as sorbent for magnetic solid phase extraction coupled with high performance liquid chromatography for simultaneous extraction and determination of tricyclic antidepressants. *Anal. Chim. Acta* **2018**, *1034*, 204–213. [CrossRef]
27. Hamidi, F.; Hadjmohammadi, M.R.; Aghaie, A.B.G. Ultrasound-assisted dispersive magnetic solid phase extraction based on amino-functionalized Fe_3O_4 adsorbent for recovery of clomipramine from human plasma and its determination by high performance liquid chromatography: Optimization by experimental design. *J. Chromatogr. B* **2017**, *1063*, 18–24. [CrossRef]
28. Ahmadi, F.; Mahmoudi-Yamchi, T.; Azizian, H. Super paramagnetic core-shells anchored onto silica grafted with C_8/NH_2 nano-particles for ultrasound-assisted magnetic solid phase extraction of imipramine and desipramine from plasma. *J. Chrom. B* **2018**, *1077–1078*, 52–59. [CrossRef]
29. Scoggins, B.A.; Maguire, K.P.; Norman, T.R.; Burrows, G.D. Measurement of tricyclic antidepressants. Part. I. A review of methodology. *Clin. Chem.* **1980**, *26*, 5–17.
30. Gupta, R.N.; Stefanec, M.; Eng, F. Determination of tricyclic antidepressant drugs by gas chromatography with the use of a capillary column. *Clin. Biochem.* **1983**, *2*, 94–97. [CrossRef]
31. Van Brunt, N. Application of new technology for the measurement of tricyclic antidepressants using capillary gas chromatography with a fused silica DB5 column and nitrogen phosphorus detection. *Ther. Drug Monit.* **1983**, *5*, 11–37. [PubMed]
32. Norman, T.R.; Maguire, K.P. Analysis of tricyclic antidepressant drugs in plasma and serum by chromatographic techniques. *J. Chromatogr.* **1985**, *340*, 173–197. [CrossRef]
33. Smyth, W.F.J. Recent studies on the electrospray ionisation mass spectrometric behavior of selected nitrogen-containing drug molecules and its application to drug analysis using liquid chromatography-electrospray ionisation mass spectrometry. *J. Chromatogr. B* **2005**, *824*, 1–20. [CrossRef] [PubMed]
34. Maurer, H.H. Multi-analyte procedures for screening for and quantification of drugs in blood, plasma, or serum by liquid chromatography-single stage or tandem mass spectrometry (LC-MS or LC-MS/MS) relevant to clinical and forensic toxicology. *Clin. Biochem.* **2005**, *38*, 310–318. [CrossRef] [PubMed]
35. Curry, S.H. Determination of Nanogram Quantities of Chlorpromazine and Some of Its metabolites in Plasma Using Gas-Liquid Chromatography with an Electron Capture Detector. *Anal. Chem.* **1968**, *40*, 1251–1255. [CrossRef]
36. Gifford, A.; Turner, P.; Pare, C.M.B. Sensitive method for the routine determination of tricyclic antidepressants in plasma using a specific nitrogen detector. *J. Chromatogr.* **1975**, *105*, 107–113. [CrossRef]
37. Jorgensen, A. Gas Chromatographic Method for the Determination of Amitriptyline and Nortriptyline in Human Serum. *Acta Pharmacol. Toxicol.* **1975**, *36*, 79–90. [CrossRef]

38. Vasiliades, J.; Buch, K.C. Gas Liquid Chromatographic Determination of Therapeutic and Toxic Levels of Amitriptyline in Human Serum with a Nitrogen-Sensitive Detector. *Anal. Chem.* **1976**, *48*, 1708–1711. [CrossRef]
39. Bailey, D.N.; Jatlow, P.I. Gas-Chromatographic Analysis for Therapeutic Concentrations of lmiprarnine and Desipramine in Plasma, with Use of a Nitrogen Detector. *Clin. Chem.* **1976**, *22*, 1697–1701.
40. Claeys, M.; Muscettola, G.; Markey, S.P. Simultaneous measurement of imipramine and desipramine by selected ion recording with deuterated internal standards. *Biomed. Mass Spectrom.* **1976**, *3*, 110–116. [CrossRef]
41. Dorrity, F.; Linnolla, M.; Habig, R.L. Therapeutic Monitoringo f Tricyclic Antidepressantsin Plasma by Gas Chromatography. *Clin. Chem.* **1977**, *23*, 1326–1328. [PubMed]
42. Wilson, J.M.; Williamson, L.J.; Raisys, V.A. Simultaneous Measurement of Secondary and Tertiary Tricyclic Antidepressants by GC/MS Chemical Ionization Mass Fragmentography. *Clin. Chem.* **1977**, *23*, 1012–1017.
43. Garland, W.A. Quantitative determination of amitriptyline and its principal metabolite, nortriptyline, by GLC-chemical ionization mass spectrometry. *J. Pharm. Sci.* **1977**, *1*, 77–81. [CrossRef]
44. Garland, W.A.; Muccino, R.R.; Min, B.H.; Cupano, J.; Fann, W.E. A method for the determination of amitriptyline and its metabolites nortriptyline, 10-hydroxyamitriptyline, and 10-hydroxynortriptyline in human plasma using stable isotope dilution and gas chromatography-chemical ionization mass spectrometry (GC-CIMS). *Clin Pharmacol. Ther.* **1979**, *6*, 844–856. [CrossRef]
45. Dhar, A.K.; Kutt, H. An improved gas-liquid chromatographic procedure for the determination of amitriptyline and nortriptyline levels in plasma using nitrogen-sensitive detectors. *Ther. Drug Monit.* **1979**, *31*, 209–216. [CrossRef]
46. Abernethy, D.R.; Greenblatt, D.J.; Shader, R.I. Tricyclic Antidepressants Determination in Human Plasma by Gas-Liquid Chromatgraphy Using Nitrogen-Phosporous Detection: Application to Single-Dose Pharmacokinetic Studies. *Pharmacology* **1981**, *23*, 57–63. [CrossRef] [PubMed]
47. Narasimhachari, N.; Saady, J.; Friedel, R.O. Quantitative mapping of metabolites of imipramine and desipramine in plasma samples by gas chromatographic-mass spectrometry. *Biol. Psychiatry* **1981**, *10*, 937–944.
48. Hals, P.A.; Lundgren, T.I.; Aarbakke, J. A sensitive gas chromatographic assay for amitriptyline and nortriptyline in plasma. *Ther. Drug Monit.* **1982**, *4*, 365–369. [CrossRef] [PubMed]
49. Jones, D.R.; Lukey, B.J.; Hurst, H.E. Quantification of amitriptyline, nortriptyline, and 10-hydroxy metabolite isomers in plasma by capillary gas chromatography with nitrogen-sensitive detection. *J. Chromatogr.* **1983**, *278*, 291–299. [CrossRef]
50. Ishida, R.; Ozaki, T.; Uchida, H.; Irikura, T. Gas chromatographic–mass spectrometric determination of amitriptyline and its major metabolites in human serum. *J. Chromatogr.* **1984**, *305*, 73–82. [CrossRef]
51. Hattori, H.; Takashima, E.; Yamada, T. Detection of tricyclic antidepressants in body fluids by gas chromatography with a surface ionization detector. *J. Chromatogr.* **1990**, *529*, 189–193. [CrossRef]
52. Ulrich, S.; Isensee, T.; Pester, U. Simultaneous determination of amitriptyline, nortriptyline and four hydroxylated metabolites in serum by capillary gas-liquid chromatography with nitrogen-phosphorus-selective detection. *J. Chromatogr. B Biomed. Appl.* **1996**, *685*, 81–89. [CrossRef]
53. Pommier, F.; Sioufli, A.; Godbillon, J. Simultaneous determination of imipramine and its metabolite desipramine in human plasma by capillary gas chromatography with mass-selective detection. *J. Chromatogr. B* **1997**, *703*, 147–158. [CrossRef]
54. Lee, X.; Kumazawa, T.; Sato, K. Detection of Tricyclic Antidepressants in Whole Blood by Headspace Solid-Phase Microextraction and Capillary Gas Chromatography. *J. Chromatogr. Sci.* **1997**, *37*, 302–308. [CrossRef]
55. De la Torre, R.; Ortuño, J.; Pascual, J.A.; González, S.; Ballesta, J. Quantitative Determination of Tricyclic Antidepressants and Their Metabolites in Plasma by Solid-Phase Extraction (Bond- Elut TCA) and Separation by Capillary Gas Chromatography with Nitrogen- Phosphorous Detection. *Ther. Drug Monit.* **1998**, *20*, 340–346. [CrossRef] [PubMed]
56. Way, B.A.; Stickle, D.; Mitchell, M.E.; Koenig, J.W.; Turk, J. Isotope Dilution Gas Chromatographic- Mass Spectrometric Measurement of Tricyclic Antidepressant Drugs. Utility of the 4-Carbethoxyhexafluorobutyryl Derivatives of Secondary Amines. *J. Anal. Toxicol.* **1998**, *22*, 374–382. [CrossRef]

57. Mellstrom, B.; Eksborg, S. Determination of chlorimipramine and desmethylchlorimipramine in human plasma by ion-pair partition chromatography. *J. Chromatogr.* **1976**, *116*, 475–479. [CrossRef]
58. Paterson, S.; Cordero, R.; Burlinson, S. Screening and semi-quantitative analysis of post mortem blood for basic drugs using gas chromatography/ion trap mass spectrometry. *J. Chromatogr. B* **2004**, *813*, 323–330. [CrossRef] [PubMed]
59. Crifasi, J.A.; Bruder, M.F.; Long, C.W.; Janssen, K. Performance Evaluation of Thermal Desorption System (TDS) for Detection of Basic Drugs in Forensic Samples by GC-MS. *J. Anal. Toxicol.* **2006**, *30*, 582–592. [CrossRef]
60. Sarafraz-Yazdi, A.; Yazdinejad, S.R.; Es'haghi, Z. Directly Suspended Droplet Microextraction and Analysis of Amitriptyline and Nortriptyline by GC. *Chromatographia* **2007**, *66*, 613–617. [CrossRef]
61. Rana, S.; Uralets, V.P.; Ross, W. A New Method for Simultaneous Determination of Cyclic Antidepressants and their Metabolites in Urine Using Enzymatic Hydrolysis and Fast GC-MS. *J. Anal. Toxicol.* **2008**, *32*, 355–363. [CrossRef]
62. Lee, X.; Hasegawa, C.; Kumazawa, T.; Shinmen, N.; Shoji, Y.; Seno, H.; Sato, K. Determination of tricyclic antidepressants in human plasma using pipette tip solid-phase extraction and gas chromatography–mass spectrometry. *J. Sep. Sci.* **2008**, *31*, 2265–2271. [CrossRef] [PubMed]
63. Ito, R.; Ushiro, M.; Takahashi, Y.; Saito, K.; Ookubo, T.; Iwasaki, Y.; Nakazawa, H. Improvement and validation the method using dispersive liquid–liquid microextraction with in situ derivatization followed by gas chromatography–mass spectrometry for determination of tricyclic antidepressants in human urine samples. *J. Chromatogr. B* **2011**, *879*, 3714–3720. [CrossRef] [PubMed]
64. Rani, S.; Kumar, A.; Malik, A.K.; Singh, B. Quantification of Tricyclic and Nontricyclic Antidepressants in Spiked Plasma and Urine Samples Using Microextraction in Packed Syringe and Analysis by LC and GC-MS. *Chromatographia* **2011**, *74*, 235–242. [CrossRef]
65. Papoutsis, I.; Khraiwesh, A.; Nikolaou, P.; Pistos, C.; Spiliopoulou, C.; Athanaselis, S. A fully validated method for the simultaneous determination of 11 antidepressant drugs in whole blood by gas chromatography–mass spectrometry. *J. Pharm. Biomed. Anal.* **2012**, *70*, 557–562. [CrossRef] [PubMed]
66. Farag, R.S.; Darwish, M.Z.; Hammad, H.A.; Fathy, W.M. Validated method for the simultaneous determination of some tricyclic antidepressants in human urine samples by gas chromatography–mass spectrometry. *Int. J. Anal. Bioanal. Chem.* **2013**, *3*, 59–63.
67. Dos Santos, M.F.; Ferri, C.C.; Seulin, S.C.; Leyton, V.; Pasqualucci, C.A.G.; Munoz, D.R.; Yonamine, M. Determination of antidepressants in whole blood using hollow-fiber liquid-phase microextraction and gas chromatography–mass spectrometry. *Forensic Toxicol.* **2014**, *32*, 214–224. [CrossRef]
68. Banitaba, M.H.; Davarani, S.S.H.; Ahmar, H.; Movahed, S.K. Application of a new fiber coating based on electrochemically reduced graphene oxide for the cold-fiber headspace solid-phase microextraction of tricyclic antidepressants. *J. Sep. Sci.* **2014**, *37*, 1162–1169. [CrossRef]
69. Mohebbi, A.; Yaripour, S.; Farajzadeha, M.A.; Mogaddamd, M.R.A. Combination of dispersive solid phase extraction and deep eutectic solvent–based air-assisted liquid–liquid microextraction followed by gas chromatography–mass spectrometry as an efficient analytical method for the quantification of some tricyclic antidepressant drugs in biological fluids. *J. Chromatogr. A* **2018**, *1571*, 84–93. [CrossRef]
70. Jagtap, P.K.; Tapadia, K. Pharmacokinetic determination and analysis of nortriptyline based on GC–MS coupled with hollow-fiber drop-to-drop solvent microextraction technique. *Bioanalysis* **2018**, *10*, 143–152. [CrossRef]
71. Martinez, M.A.; de la Torre, C.S.; Almarza, E. Simultaneous Determination of Viloxazine, Venlafaxine, Imipramine, Desipramine, Sertraline, and Amoxapine in Whole Blood: Comparison of Two Extraction/Cleanup Procedures for Capillary Gas Chromatography with Nitrogen-Phosphorus Detection. *J. Anal. Toxicol.* **2002**, *26*, 296–302. [CrossRef]
72. Martinez, M.A.; de la Torre, C.S.; Almarza, E. A Comparative Solid-Phase Extraction Study for the Simultaneous Determination of Fluoxetine, Amitriptyline, Nortriptyline, Trimipramine, Maprotiline, Clomipramine and Trazodone in Whole Blood by Capillary Gas-Liquid Chromatography with Nitrogen-Phosphorus Detection. *J. Anal. Toxicol.* **2003**, *27*, 353–358. [CrossRef]
73. Yazdi, A.S.; Razavi, N.; Yazdinejad, S.R. Separation and determination of amitriptyline and nortriptyline by dispersive liquid–liquid microextraction combined with gas chromatography flame ionization detection. *Talanta* **2008**, *75*, 1293–1299. [CrossRef]

74. Rezazadeh, M.; Yamini, Y.; Seidi, S.; Ebrahimpour, B. Electromembrane surrounded solid phase microextraction: A novel approach for efficient extraction from complicated matrices. *J. Chromatogr. A* **2013**, *1280*, 16–22. [CrossRef]
75. Yazdi, A.S.; Razavi, N. Separation and Determination of Amitriptyline and Nortriptyline in Biological Samples Using Single-Drop Microextraction with GC. *Chromatographia* **2011**, *73*, 549–557. [CrossRef]
76. Jeannot, M.A.; Cantwell, F.F. Solvent microextraction into a single drop. *Anal. Chem.* **1996**, *68*, 2236–2240. [CrossRef]
77. Davarani, S.S.H.; Najarian, A.M.; Nojavan, S.; Tabatabaei, M. Electromembrane extraction combined with gas chromatography for quantification of tricyclic antidepressants in human body fluid. *Anal. Chim. Acta* **2012**, *725*, 51–56. [CrossRef]
78. Saraji, M.; Mehrafza, N.; Hajialiakbari, A.A.; Mohammad, B.; Jafari, T. Determination of desipramine in biological samples using liquid–liquid–liquid microextraction combined with in-syringe derivatization, gas chromatography, and nitrogen/phosphorus detection. *J. Sep. Sci.* **2012**, *35*, 2637–2644. [CrossRef]
79. Davarani, S.S.H.; Nojavan, S.; Asadi, R.; Banitaba, H.M. Electro-assisted solid-phase microextraction based on poly(3,4-etylenedioxythiophen) combined with GC for the quantification of tricyclic antidepressants. *J. Sep. Sci.* **2013**, *36*, 2315–2322. [CrossRef]
80. Seidi, S.; Yamini, Y.; Rezazadeh, M. Combination of electromembrane extraction with dispersive liquid–liquid microextraction followed by gas chromatographic analysis as a fast and sensitive technique for determination of tricyclic antidepressants. *J. Chromatogr. B* **2013**, *913–914*, 138–146. [CrossRef]
81. Asghari, A.; Saffarzadeh, Z.; Bazregar, M.; Rajabi, M.; Boutorabi, L. Low-toxic air-agitated liquid-liquid microextraction using a solidifiable organic solvent followed by gas chromatography for analysis of amitriptyline and imipramine in human plasma and wastewater samples. *Microchem. J.* **2017**, *130*, 122–128. [CrossRef]
82. Karageorgou, E.; Manousi, N.; Samanidou, V.F.; Kabir, A.; Furton, K.G. Fabric Phase Sorptive Extraction for the Fast Isolation of Sulfonamides Residues from Raw Milk Followed by High Performance Liquid Chromatography with Ultraviolet Detection. *Food Chem.* **2016**, *196*, 428–436. [CrossRef] [PubMed]
83. Samanidou, V.F.; Georgiadis, D.; Kabir, A.; Furton, K.G. Capsule Phase Microextraction: The Total and Ultimate Sample Preparation Approach. *J. Chromogr. Sep. Tech* **2018**, *9*, 1–4. [CrossRef]

© 2019 by the authors. Licensee MDPI, Basel, Switzerland. This article is an open access article distributed under the terms and conditions of the Creative Commons Attribution (CC BY) license (http://creativecommons.org/licenses/by/4.0/).

Article

Insights into the Mechanism of Separation of Bisphosphonates by Zwitterionic Hydrophilic Interaction Liquid Chromatography: Application to the Quantitation of Risedronate in Pharmaceuticals

Irene Panderi [1,*], Eugenia Taxiarchi [1], Constantinos Pistos [2], Eleni Kalogria [1] and Ariadni Vonaparti [3]

1. National and Kapodistrian University of Athens, School of Pharmacy, Division of Pharmaceutical Chemistry, Laboratory of Pharmaceutical Analysis, Panepistimiopolis, Zografou, 157 71 Athens, Greece; eugenetax@live.com (E.T); ekalogria@yahoo.gr (E.K.)
2. West Chester University, Department of Chemistry, West Chester, PA 19383, USA; cpistos@wcupa.edu
3. Qatar Doping Analysis Laboratory, Sports City Road, Aspire Zone, PO Box 27775 Doha, Qatar; avonaparti@adlqatar.qa
* Correspondence: irenepanderi@gmail.com; Tel.: +30-697-401-5798

Received: 2 January 2019; Accepted: 17 January 2019; Published: 22 January 2019

Abstract: Bisphosphonates are used to treat various skeletal disorders, as they modulate bone metabolism by inhibition of the osteoclast-mediated bone resorption. These compounds are both polar and ionic, and therefore, by using reversed phase liquid chromatography are eluted rapidly. Hydrophilic interaction liquid chromatography (HILIC) is an advantageous technique for the separation and analysis of polar molecules. As the elution order in HILIC is reversed to reversed phase liquid chromatography, a reasonable retention and selectivity for polar compounds is expected. In this work the retention mechanism of three bisphosponates, namely risedronate, tiludronate and zoledronate, was investigated under zwitterionic HILIC conditions. The key factors influencing the retention of the analytes on a zwitterionic ZIC®-pHILIC column (150.0 × 2.1 mm i.d., 200 Å, 3.5 µm) have been systematically investigated. It was found that apart from partition, electrostatic repulsions play an important role in the retention of bisphosphonates. Peak tailing of risedronate and zoledronate was improved by the addition of sodium pyrophosphate in the mobile phase. A zwitterionic hydrophilic interaction liquid chromatography-photodiode array (HILIC-PDA) method was further optimized and fully validated to quantitate risedronate in commercial film-coated tablets. The calibration curves for risedronate showed good linearity ($r \geq 0.9991$) within the calibration range tested. The intra- and inter-day coefficient of variation (CV) values was less than 0.6%, while the relative percentage error (%Er) was less than −2.3%. Accelerated stability studies of risedronate conducted under several degradation conditions including hydrolysis, oxidation and heat demonstrated the selectivity of the procedure. A short-run analysis of not more than 6 min allowed the analysis of large samples per day. The applicability of the method for the quantitation of risedronate was demonstrated via the analysis of commercial tablets containing this compound.

Keywords: bisphosphonates; risedronate; zoledronate; tiludronate; ZIC-HILIC; PDA; quantitation; tablets

1. Introduction

Bisphosphonates belong to a unique class of drugs which are chemically stable analogues of the inorganic pyrophosphate anion, a secondary product of various biochemical processes. The concentration levels of pyrophosphate anion in blood are associated with the mechanism of bone calcification [1]. Like pyrophosphate, bisphosphonates have high affinity for bone mineral and bind strongly to hydroxyapatite calcium in the bone. The skeletal accumulation of bisphosphonates (on the skeleton) depends highly on the disposal of hydroxyapatite binding sites. Non-nitrogen containing bisphosphonates are accumulated into newly formed adenosine triphosphate (ATP) analogues and inhibit ATP-dependent processes, leading to osteoclast apoptosis [2]. Conversely, nitrogen containing bisphosphonates inhibit the action of the enzyme farnesyl pyrophosphate synthase (FPPS) enzyme, which is involved in the mevalonate pathway [3]. These drugs have become the therapy of choice for the management of various skeletal disorders such as several types of osteoporosis, hypercalcemia, Paget disease and malignancy metastatic to bone [4]. However, despite the well-recognized benefits of bisphosphonates, these drugs may cause also osteonecrosis of the jaw [5,6].

Bisphosphonates are both polar and ionic compounds and by using reversed phase liquid chromatography are eluted rapidly. The development of a chromatographic method for the analysis of bisphosphonates is a challenge for the analysts due to their high hydrophilicity. In addition to this, the lack of chromophores in some bisphosphonates structures necessitates the use of tedious and time-consuming derivatization procedures for their detection. A literature survey revealed that in most publications adequate retention of bisphosphonates is achieved by using ion-pairing agents in the mobile phase [7–10], anion-exchange chromatography [11] and in some cases pre-column [12] or post-column [13] derivatization procedures [14]. A fused core Ascentis Express HILIC column has been used to quantitate risedronate sodium in pharmaceuticals with PDA and tandem mass spectrometric detection [15]. Bisphosphonates in biological matrices have been quantified after their methylation with trimethylsilyl diazomethane by the use of liquid chromatography–mass spectrometric methods [16–21].

During the last twenty years, HILIC has been proved to be a promising technique for the analysis of polar substances. The separation mechanism in HILIC involves multiple factors, such as partitioning, normal phase/adsorption interactions, hydrogen bonding, reversed-phase and electrostatic interactions [22]. The significance of each of these mechanisms depends on the type of mobile phase and stationary phase that will be used. HILIC requires the use of highly organic mobile phases that contain an aprotic solvent (mainly acetonitrile) in combination with at least 3% of an aqueous salt solution and polar stationary phases so to facilitate chromatographic separation [23]. The major factors affecting retention in HILIC are the type of the stationary phase, the percentage content of water which is the strongest eluent, along with the concentration, pH and type of the aqueous solution of the salt [24]. Studies on the retention mechanism of compounds in HILIC are of great interest, since it is difficult to predict the effect of the operation parameters on the retention of substances. Bisphosphonates, due to their increased polarity, are perfect candidates to study their retention mechanism in HILIC. Up to now, only a limited number of publications have been reported for the analysis of bisphosphonates in HILIC [15]. This paper describes studies on the retention mechanism of two nitrogen-containing bisphosphonates, namely risedronate and zoledronate, and one non-nitrogen-containing bisphosphonate, namely tiludronate, on a polymeric zwitterion ZIC®-pHILIC column. It is the first time that zwitterionic hydrophilic interaction liquid chromatography is used to study the retention of bisphosphonates. The zwitterionic hydrophilic interaction liquid chromatography used in this work is a unique form of HILIC, which involved the use of substrates containing zwitterionic functional groups. The key factors influencing the chromatography of these analytes were systematically investigated. A HILIC stability-indicating assay method coupled with photodiode array detection was further optimized and validated to quantitate risedronate in commercial film-coated tablets. Accelerated stability studies of risedronate were also conducted under stress conditions to demonstrate the selectivity of the procedure. The applicability of the method for the quantitation of risedronate was finally proven via the analysis of commercial film-coated tablets containing risedronate as the active ingredient.

2. Materials and Methods

2.1. Chemicals and Reagents

Risedronate sodium salt, hydroxy-(1-hydroxy-1-phosphono-2-pyridin-3-ylethyl)phosphinate; sodium and tiludronate disodium salt, [(4-chlorophenyl)sulfanyl-[hydroxy(oxido) phosphoryl]methyl]-hydroxyphosphinate; disodium, were obtained from Sigma-Aldrich, Germany. Zoledronic acid monohydrate, (1-hydroxy-2-imidazol-1-yl-1-phosphonoethyl)phosphonic acid; hydrate, was obtained from TGI Tokyo Chemical Industry Co., Ltd (Tokyo, Japan). All bisphosphonates were of pharmaceutical purity grade. Solvents of HPLC grade were obtained from E. Merck, Germany. Ammonium formate, ammonium acetate, acetic acid and sodium pyrophosphate were purchased from Acros Organics part of Thermo Fischer Scientific (Geel, Belgium). Water was deionized and further purified by means of a Merck Millipore Synergy UV system (Darmstadt, Germany). Kinesis KX hydrophilic polytetrafluoroethylene (PTFE) syringe filters (diameter 13 mm, pore size 0.22 µm) were purchased from Kinesis Ltd, Cambridge shire, UK.

Commercial film-coated tablet labelled to contain 35 mg risedronate sodium (equivalent to 32.5 mg risedronic acid). Inactive ingredients of the tablet core consist of crospovidone A, cellulose microcrystalline, magnesium stearate and lactose monohydrate. Inactive ingredients of the film coating consist of hypromellose, titanium dioxide E171, hydroxypropyl cellulose, macrogol, iron oxide red (E172), colloidal anhydrous silica and iron oxide yellow (E172).

2.2. Instrumentation

Experiments were performed on an HPLC-PDA system consisting of an auto sampler model Waters 717 plus, a constant temperature oven, an isocratic pump model Waters 1515 and a photodiode array detector model Waters 996 (Milford, MA, USA). Data acquisition and analysis was attained by the use of the Empower software (Milford, MA, USA). The analytes were detected over the wavelength range of 200 to 400 nm and the chromatograms were extracted at $\lambda = 262$ nm. Chromatography was performed by using a polymeric zwitterionic ZIC®-pHILIC analytical column (150.0 × 2.1 mm i.d., 200 Å, particle size 3.5 µm) (Merck Millipore, Darmstadt, Germany). Moreover, a guard column (20 × 2.1 mm, 3.5 µm) of the same packing material was used to prolong the column lifetime. During the method development, various mobile phases consisting of mixtures of acetonitrile and ammonium formate or ammonium acetate aqueous solutions were used and the flow rate was set at 0.25 mL min^{-1}. Aqueous solutions of ammonium acetate were prepared freshly every day. For the quantitation of risedronate, the mobile phase consisted of 38% 9 mM ammonium acetate and 1 mM sodium pyrophosphate aqueous solution pH 8.8 in acetonitrile and pumped at a flow rate of 0.15 mL min^{-1}. It was filtered through a 0.22 µm Nylon-membrane filter, Membrane Solutions (Kent, WA, USA) and degassed under vacuum prior to use. Chromatography was performed at 40 ± 2 °C, with a chromatographic run time of 6 min; a 60 µL volume was injected into a 10 µL loop.

2.3. Statistical Analysis

Regression analysis was performed using IBM SPSS Statistics ver. 22, IBM software. The ionization state of each compound was estimated using ADME boxes ver. 3.0, Pharma Algorithms software.

2.4. Stock and Working Standard Solutions

Stock standard solutions of risedronate, zoledronate and tiludronate were prepared at 500 µg mL^{-1} in acetonitrile-water mixture (60:40, v/v). Stock standard solution of risedronate was prepared in duplicate for the calibration standards and the quality control samples. These solutions were stable for several weeks when stored at −17 °C for several months. The stock standard solutions were further diluted in acetonitrile to prepare working standard solutions at two concentration levels 5 and 10 µg mL^{-1} for each analyte. These solutions were used for the method development and were stored under refrigeration at 4 °C for two months.

Calibration standard solutions of risedronate were prepared in acetonitrile over the concentration range of 1.5 to 5 µg mL^{-1}. Quality control samples of risedronate were also prepared in acetonitrile at three concentration levels (1.5, 3.5 and 5 µg mL^{-1}). Calibration standard solutions and quality control samples were prepared freshly every day and remained stable throughout the analysis.

2.5. Assay Procedure for the Pharmaceutical Samples

To calculate the tablet weight, 10 tablets containing 35 mg of risedronate sodium were weighted and then pulverized. A portion of this powder, equivalent to 35 mg of risedronate sodium, was transferred into a 100 mL volumetric flask and diluted to volume with acetonitrile/water mixture (10:90, v/v). The mixture was sonicated for 10 min and then transferred into a 2 mL Eppendorf tube for centrifugation at 4.000× g and 25 °C for 10 min. The supernatant was then sonicated in an ultrasonic bath for additional 10 min and filtered through a PTFE hydrophilic syringe filter. A 100 µL aliquot of the filtrate was then transferred into a 10 mL volumetric flask and diluted to volume with acetonitrile prior to HILIC-PDA analysis.

2.6. Accelerated and Long-Term Stability Studies

Degradation studies were performed in risedronate under various stress conditions where degradation was stimulated by acidic or basic hydrolysis, oxidation and thermal degradation. Risedronate bulk substance was stressed under accelerated degradation conditions with 1.0 M HCl at 50 °C (± 2) for 10 days, 1.0 M potassium hydroxide (NaOH) at 50 °C (± 2) for 24 h and 3.0% v/v hydrogen peroxide (H_2O_2) at 25 °C (± 2) for 3 h. The concentration of risedronate bulk substance in the accelerated stability samples was 0.35 mg mL^{-1}. During each degradation experiment and at predetermined time intervals, appropriate aliquots were neutralized with base or acid, and analyzed according to the proposed method. The concentration of risedronate in the analyzed sample solution was 3.5 µg mL^{-1}.

Blistered tablets containing risedronate were stressed in long-term stability studies. Blistered tablets have been stored for 3 months at 50 °C (± 2) and 75% (± 2) relative humidity, and at 50 °C ± 2 °C and 15% (± 2) relative humidity. After the completion of each degradation treatment the samples were analyzed as described in the sample preparation procedure (Section 2.5).

3. Results and Discussion

3.1. Method Development

Bisphosphonates contain two phosphoric acid groups and are strongly polar and ionic compounds. The three bisphosphonates drugs studied in this work are divided into two groups: two nitrogen-containing compounds (risedronate and zoledronate) and one acidic compound (tiludronate). These compounds are poorly retained in the classical reversed phase analytical columns and their chromatographic analysis is challenging. The ZIC®-pHILIC analytical column used is a polymeric and zwitterionic sulfoalkylbetaine stationary phase. The functional group of this column consists of a sulfonic acid group (acidic), which was separated with a short alkyl spacer from a quaternary ammonium group (basic). In this zwitterionic stationary phase, the electrostatic forces of each charge were partly counterbalanced by the proximity of an ion with opposite charge. Though the accessibility to the positively charged quaternary ammonium groups was limited, the negatively charged sulfonic acid groups might be responsible for weak, but important, electrostatic interactions [25]. The studied bisphosphonates were retained adequately in this analytical column through hydrophilic interactions, even if they had the same charge with the sulfonic acid groups of the stationary phase. Electrostatic repulsions in HILIC were first described by Alpert [26] as electrostatic repulsion hydrophilic interaction chromatography (ERLIC). These kinds of interactions were of great interest and can be used to selectively antagonize the retention of analytes that normally would be best retained [27].

3.1.1. Effect of Chromatographic Parameters on the Bisphosphonates Retention

A one-variable-at-a-time approach was used to study the chromatography of bisphosphonates in the zwitterionic stationary phase. Mobile phases in HILIC typically contain high percentages of acetonitrile mixed with an aqueous salt solution. In this work, the mobile phase salts were limited to ammonium formate and ammonium acetate due to their good solubility in acetonitrile. The sulfonic acid groups of the stationary phase are responsible for weak electrostatic interactions that can be reduced by the addition of an aqueous salt solution. In preliminary experiments with a mobile phase containing 35% 10 mM ammonium formate water solution in acetonitrile, both nitrogen-containing bisphosphonates (risedronate and zoledronate) were not eluted, while tiludronate exhibited a broad asymmetrical peak. On the other hand, ammonium acetate improved the chromatography for all compounds. Consequently, ammonium acetate concentration was varied from 1 to 40 mM in mobile phases containing 35% Φ_{water}. The logarithm of retention factor (k) was used to evaluate retention of the analytes. Retention factor (k) is independent of column geometry and flow rate and was often used for reproducibility evaluation, and method validation [28].

Typical HILIC chromatograms illustrating the effect of the concentration of ammonium acetate on the retention time and the peak shape bisphosphonates are presented in Figure 1. In all of the ammonium acetate concentrations tested tiludronate exhibits good peak symmetry while tailing peaks are observed for both risedronate and zoledronate. Bisphosphonates as strong chelators are capable to interact with the metals of the liquid chromatographic (LC) system [10,29]. This binding affinity is greater in nitrogen-containing bisphosphonates where the one side chain of the molecule is a primary amino-group and allows a tridentate interaction [30]. By increasing ammonium acetate concentration, the elution of the analytes was delayed, leaving them more time to interact with the metals of the LC system; hence, peak tailing of nitrogen-containing bisphosphonates increased.

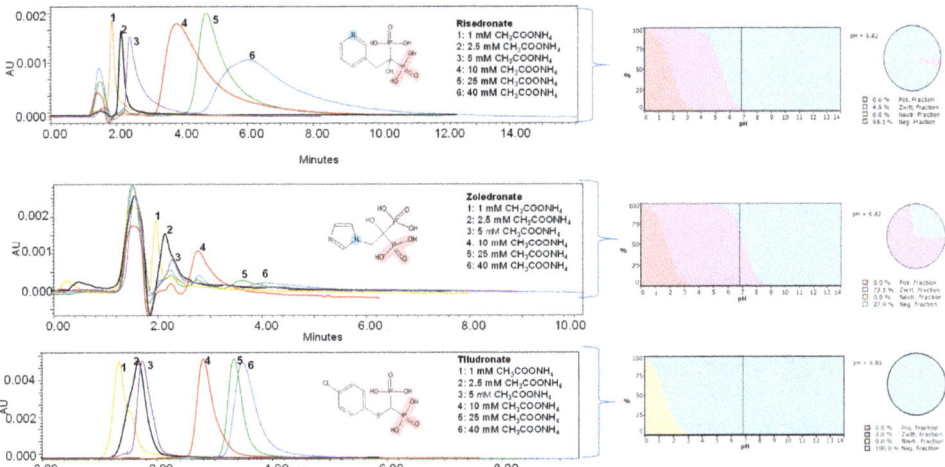

Figure 1. Typical HILIC chromatograms displaying the effect of ammonium acetate concentration on the peak shape and the retention of bisphosphonates, accompanied by diagrams of their ionization state at pH 6.8 as calculated ADME boxes ver. 3.0, Pharma Algorithms software. Chromatographic conditions: ZIC®-pHILIC analytical column, mobile phase: aqueous solution of ammonium acetate pH 6.8/acetonitrile (35:65, v/v), 0.25 mL min^{-1} flow rate and wavelength of detection at 262 nm.

As illustrated in Figure 2A, the retention of both the nitrogen-containing bisphosphonates (risedronate and zoledronate) and the negatively charged tiludronate increased by increasing ammonium acetate concentration due to the reduction of the electrostatic repulsions between the analytes and the stationary phase. From these experiments, we concluded that by using a 10 mM ammonium acetate concentration all bisphosphonates are adequately retained and well separated from the solvent front.

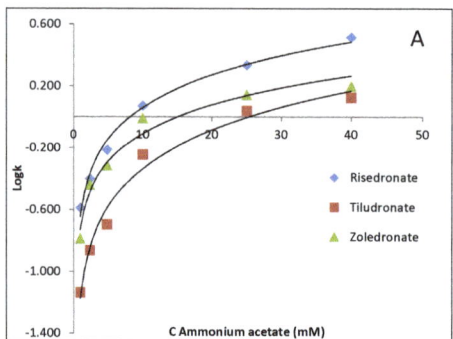 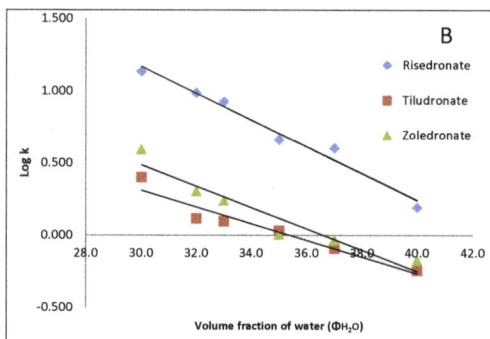

Figure 2. (**A**) Impact of the concentration of ammonium acetate (mM) on the log k. ZIC®-pHILIC column; mobile phase: acetonitrile/ammonium acetate aqueous solution pH 6.6 (65:35, v/v), (**B**) Impact of the percentage of water, Φ_{water}, on the log k. ZIC®-pHILIC column; mobile phase: acetonitrile/ammonium acetate aqueous solution pH 6.8 containing 3.5 mM ammonium acetate in whole monile phase.

In HILIC a minimum percentage of water, Φ_{water}, at 2% to 3% in the mobile phase is crucial for the creation of the water layer around the stationary phase. In mobile phases with high percentages of acetonitrile, the elution of polar compounds was increased, since the water interacts strongly with the polar stationary phase. To study the effect of Φ_{water} on the retention of bisphosphonates, the concentration of ammonium acetate in whole mobile phase stayed constant at 3.5 mM, while Φ_{water} varied from 30% to 40 %. As shown in Figure 2B the retention of all analytes decreases linearly with increasing Φ_{water}, implying partition as the dominant retention mechanism for bisphosphonates in HILIC.

The studies presented above indicate that both hydrophilic partition and secondary electrostatic interactions contribute to the retention of bisphosphonates on the ZIC®-pHILIC analytical column. Bisphosphonates are strong chelators and their interaction with the metals of the LC system causes serious peak tailing [29]. This binding affinity of bisphosphonates is greater in nitrogen-containing bisphosphonates, where the one side chain of the molecule is a primary amino-group and allows a tridentate interaction [30]. The presence of phosphate groups in the mobile phase can be critical for the analysis of bisphosphonates on a standard stainless steel LC system [31]. To overcome peak tailing for nitrogen-containing bisphosphonates, sodium pyrophosphate was added to the aqueous content of the mobile phase. As can be seen in Figure 3, risedronate and zoledronate peak tailing is seriously reduced in the presence of sodium pyrophosphate, since pyrophosphate anions interact selectively with the metals of the LC system [10]. Tiludronate retention is not seriously affected by the presence of sodium pyrophosphate in the mobile phase.

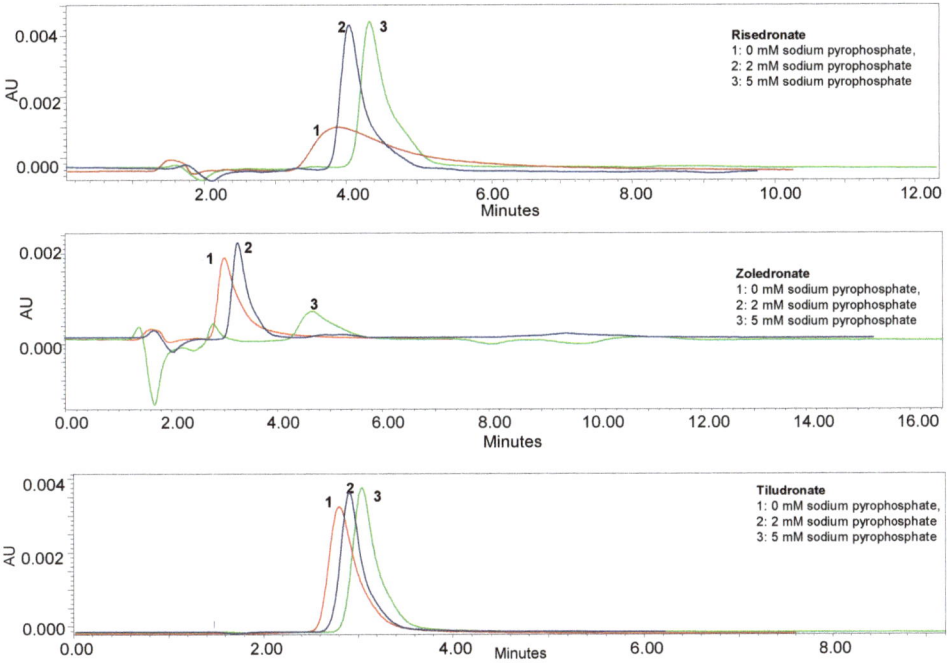

Figure 3. Typical HILIC chromatograms showing the effect of sodium pyrophosphate concentration (mM) on the on the retention time and the peak shape of bisphosphonates. Chromatographic conditions: ZIC®-pHILIC analytical column, mobile phase: acetonitrile–10 mM ammonium acetate aqueous solution pH 6.8 (65:35, v/v), flow rate of 0.25 mL min^{-1} and UV detection at 262 nm.

3.1.2. Optimization of the Chromatographic Parameters for the Quantitation of Risedronate

It was observed that the back pressure of the chromatographic system was increased by increasing sodium pyrophosphate concentration; thus, it was decided to reduce its concentration to 1 mM and to decrease the flow rate to 0.15 mL min^{-1}. Moreover, the addition of sodium pyrophosphate salt in the aqueous content of the mobile phase resulted in alkaline pH that was adjusted to 8.8 using acetic acid. By keeping sodium pyrophosphate concentration constant at 1 mM, a one-variable-at-a-time approach was used to identify the optimal mobile phase composition for the quantitation of risedronate in tablets. The parameters selected to study were the percentage of water, Φ_{water} and the concentration of ammonium acetate (mM). It was found that an increase in the percentage of water from 35% to 39% reduced the retention factor of risedronate. Moreover, it was observed that an increase in the concentration of ammonium acetate from 6 to 10 mM increased the retention of the negatively charged risedronate due to the disruption of the electrostatic repulsions between this and the negatively charged sulfonic acid groups of the stationary phase. Thus, a mobile phase consisting of 38% 9 mM ammonium acetate and 1 mM sodium pyrophosphate aqueous solution pH 8.8 in acetonitrile was finally used. At the beginning of each experiment, the column was equilibrated for 1.5 h and column temperature was set at 40 °C. Due to the isocratic separation, there was no need for time-consuming re-equilibration of the analytical column.

3.2. Method Validation

3.2.1. Selectivity

The selectivity of the proposed HILIC-PDA method is demonstrated in Figure 4, where a representative chromatogram obtained from the analysis of standard solution containing risedronate at 3.5 µg mL^{-1} is presented overlaid with a chromatogram obtained from the analysis of risedronate commercial tablets and a blank sample (dilution solvent). No significant interfering peaks have been observed at the retention time of risedronate, which is eluted at 4.59 min.

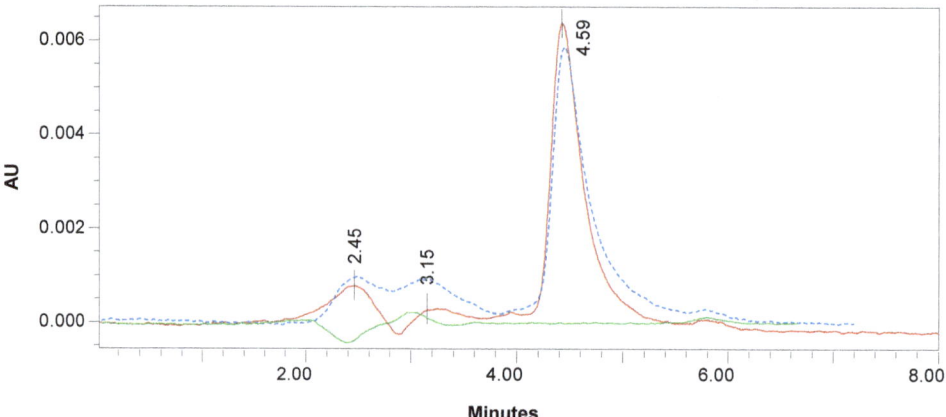

Figure 4. Typical HILIC chromatograms of a blank sample (green line) overlaid with a 3.5 µg mL^{-1} risedronate standard solution (red line) and a solution obtained from the analysis of risedronate commercial tablets (blue line). Chromatographic conditions: ZIC®-pHILIC analytical column, mobile phase: 38% 9 mM ammonium acetate and 1 mM sodium pyrophosphate aqueous solution pH 8.8 in acetonitrile, flow rate of 0.15 mL min^{-1} and UV detection at 262 nm.

3.2.2. Statistical Analysis of Data

The calibration curves of the peak area signals of risedronate versus the corresponding concentrations were linear, as shown by the results presented in Table 1. Back-calculated concentrations in the calibration curves were within 2.1% of the nominal values, which are in agreement with international guidelines.

Table 1. Statistical analysis of the calibration curves of risedronate.

Concentration Range, µg mL^{-1}	Regression Equations [a]	r [b]	Standard Deviation		S_r [c]
			Slope	Intercept	
Mean of 3 calibration curves over a period of 1 month					
1.5–5	$S_{Rsd} = 3.029 \times C_{Rsd} + 1.76$	≥0.9991	1.7×10^{-3}	3.2×10^{-2}	≤0.24

[a] Peak areas signal of risedronate, S_{Rsd} vs the corresponding concentration of risedronate, C_{Rsd}; [b] correlation coefficient; [c] standard error of the estimate.

The limit of detection (LOD) and the limit of quantitation (LOQ) were estimated experimentally by analyzing risedronate samples spiked at low concentrations. These limits were defined according to a signal-to-noise ratio (S/N) corresponding to 3:1 for the LOD and to at least 10:1 until a %CV of less than 2.5% was obtained for the LOQ. The LOD was found to be at 0.3 µg mL^{-1}, and the LOQ at 1.5 µg mL^{-1}.

Risedronate can be determined by appropriate precision and accuracy as is indicated by the intra- and inter-assay precision data that are presented in Table 2. Precision was evaluated by one-way analysis of variance. Intra-assay relative standard deviation values, %RSD, were between 0.3% and 0.6%, while inter-assay %RSD was no more than 0.6%. The overall accuracy of the method was expressed by the relative percentage error, Er%, that ranged from 2.3% to 1.8%.

Table 2. Accuracy and precision data.

Risedronate	Concentration ($\mu g\ mL^{-1}$)		
Added concentration	1.5	3.5	5
Run 1 (mean ± SD)	1.4669 ± 0.0051	3.556 ± 0.022	5.013 ± 0.046
Run 2 (mean ± SD)	1.4589 ± 0.0033	3.567 ± 0.012	5.022 ± 0.088
Run 3 (mean ± SD)	1.4715 ± 0.0044	3.5628 ± 0.0091	4.973 ± 0.012
Overall mean	1.4658	3.5618	5.0031
Intra-day CV(%) [a]	0.3	0.5	0.6
Inter-day CV(%) [a]	0.5	0.05	0.6
Overall accuracy Er% [b]	−2.3	1.8	0.1

[a] (n = 3 runs; 5 replicates per run).

During the method development, it was observed that both the volume fraction of water (Φ_{H_2O}) and the concentration of ammonium acetate affected the chromatography of risedronate. To assess method robustness, small deliberate variations were performed in the aforementioned parameters and in the wavelength of detection. Each parameter was changed at two levels (0 and 1) using a univariate approach and robustness was estimated by measuring the peak area signal. Ammonium acetate concentration was altered by 1.0 mM (range 8 to 9 mM), volume fraction of water Φ_{water} was altered by 2% (range 38% to 36%) and the wavelength of detection was altered by of 2 nm (range 262 to 264). A standard solution of risedronate at 3.5 $\mu g\ mL^{-1}$ was injected in three replicates under changes of the parameters mentioned above. The proposed method could be considered as robust, since %RSD values of the peak area signal of risedronate does not exceed 3.2% (acceptance criteria < 10%) in all of the tested conditions.

3.2.3. Accelerated and Long-Term Stability Studies

The results of the accelerated and the long-term stability studies are presented in Table 3. In the acid stressed samples and after 10 days, a 21.9% of risedronate was degraded using 1.0 M HCl at 50 °C, while no degradation products have been detected. In the base stressed samples and after 1 day, a 21.2% of risedronate was degraded using 1.0 M NaOH at 50 °C, while no degradation products were detected. In the acid stressed samples and after 3 h, a 17.3% of risedronate was degraded using 3.0% (v/v) H_2O_2 at 25 °C while the degradation products could not be detected. After the degradation of risedronate blistered tablets under low (15%) and high (75%) humidity conditions for three months, the percentage recovery of the analyte was 95.9% and 78.5%, respectively.

Table 3. Stability data for risedronate by HILIC-PDA.

Degradation Conditions/Time	Time	Concentration ($\mu g\ mL^{-1}$) (Mean ± S.D., n = 3)	% Recovery (Mean ± S.D., n = 3)	Degradation Products Retention Time (min)
1.0 M HCl, 50 °C	1 day	3.467 ± 0.038	99.0 ± 1.1	-
	2 days	3.469 ± 0.041	99.1 ± 1.2	-
	8 days	3.036 ± 0.040	86.7 ± 1.1	<3
	10 days	2.736 ± 0.032	78.1 ± 0.9	<3
1.0 M NaOH, 25 °C	1 day	2.757 ± 0.073	78.8 ± 2.1	<3
3 % v/v H_2O_2, 25 °C	1 h	3.211 ± 0.042	91.7 ± 1.2	-
	2 h	3.094 ± 0.041	88.4 ± 1.1	<3
	3 h	2.895 ± 0.047	82.7 ± 1.4	<3

Table 3. *Cont.*

Long-Term Stability Studies	Time	Amount (mg) Per Tablet (Mean ± S.D., n = 3)	% Recovery (Mean ± S.D., n = 3)	Degradation Products Retention Time (min)
50 ± 2 °C 15% humidity	1 month	32.79 ± 0.51	100.9 ± 1.5	-
	3 months	32.41 ± 0.46	99.7 ± 1.3	-
50 ± 2 °C 75% humidity	1 month	32.38 ± 0.44	99.6 ± 1.2	-
	3 months	25.22 ± 0.63	77.6 ± 1.8	<3

Based on the results presented in Table 3, the proposed HILIC-PDA method is stability-indicating, since it is able to quantitate risedronate in commercial formulations without any interference from degradation peaks.

3.3. Analysis of Commercial Tablets

The applicability of the proposed method was evaluated through the analysis of commercially available tablets containing 35 mg of risedronate sodium (equivalent to 32.5 mg risedronic acid). The analysis was performed on accurately weighted amount of the pulverized tablets. Percentage recovery was found to be 99.3 ± 0.7% of the label claim, or 32.3 ± 0.2 mg of risedronic acid per tablet (n = 10, RSD = 0.6%). Additionally, this method was used for the content-uniformity testing, in which many assays on the individual tablets are required. Percentage recovery was found to be 100.2 ± 1.2% of the label claim, or 32.6 ± 0.4 mg of risedronic acid per tablet (n = 10, RSD = 1.1%).

Table 4 presents data obtained from the analysis of real samples and indicate that the proposed HILIC-PDA method is applicable to the accurately quantitation of risedronate in commercially available tablets.

Table 4. Risedronate quantitation in a commercial formulation.

Test	Amount (mg) Per Tablet (Mean ± SD, n = 10)	% Recovery (Mean ± SD, n = 10)
Quality control	32.3 ± 0.2	99.3 ± 0.7
Content uniformity	32.6 ± 0.4	100.2 ± 1.2

4. Conclusions

HILIC is a popular chromatographic method for the analysis of hydrophilic compounds. The separation mechanism in HILIC is more complicated in regards to reversed phase HPLC due to the various kinds of interactions that rule the retention. Despite the increased number of publications in the last years, the separation mechanism in HILIC is still under investigation and many analysts are puzzled over the use of this chromatographic method [32,33]. Thus, understanding the influence of operational parameters in HILIC is crucial for an effective method development. The chemical nature of bisphosphonates, both polar and ionic, makes them perfect candidates for HILIC. In this work the chromatographic behavior of two nitrogen-containing bisphosponates, namely risedronate and zoledronate, and one non-nitrogen-containing bisphosphonate, namely tiludronate, has been thoroughly investigated under zwitterionic HILIC conditions. The results indicate that apart from partition, which is the dominant separation mechanism in HILIC, electrostatic repulsions play an important role to the retention of bisphosphonates. Peak tailing of the two nitrogen-containing bisphosphonates was improved by the addition of sodium pyrophosphate to the mobile phase. A zwitterionic hydrophilic interaction liquid chromatography method coupled to diode-array detection was developed, validated and applied to the quantitation of risedronate in pharmaceutical formulations. The proposed method is stability-indicating since it allows accurate and precise quantitation of risedronate in tablets without any interference from excipients or degradation products, and it is applicable to routine quality control of risedronate in tablets. A run time of less than 6 min ensures rapid quantitation.

Author Contributions: I.P., E.T., C.P. and A.V. participated in designing the study. E.T., E.K. and I.P. conducted the study. Data collection and analysis was performed by I.P., E.T., and A.V. The manuscript was written by I.P., C.P. and A.V.

Funding: This research received no external funding.

Acknowledgments: We would like to acknowledge George Chorafakis for proofreading the manuscript.

Conflicts of Interest: The authors have no conflict of interest.

References

1. Fleisch, H.; Russell, R.G.; Straumann, F. Effect of pyrophosphate on hydroxyapatite and its implications in calcium homeostasis. *Nature* **1966**, *212*, 901–903. [CrossRef] [PubMed]
2. Drake, M.T.; Clarke, B.L.; Khosla, S. Bisphosphonates: Mechanism of Action and Role in Clinical Practice. *Mayo Clin. Proc.* **2008**, *83*, 1032–1045. [CrossRef] [PubMed]
3. Plotkin, L.I.; Aguirre, J.I.; Kousteni, S.; Manolagas, S.C.; Bellido, T. Bisphosphonates and estrogens inhibit osteocyte apoptosis via distinct molecular mechanisms downstream of extracellular signal-regulated kinase activation. *J. Biol. Chem.* **2005**, *280*, 7317–7325. [CrossRef] [PubMed]
4. Kavanagh, K.L.; Guo, K.; Dunford, J.E.; Wu, X.; Knapp, S.; Ebetino, F.H.; Rogers, M.J.; Russell, R.G.; Oppermann, U. The molecular mechanism of nitrogen-containing bisphosphonates as antiosteoporosis drugs. *Proc. Natl. Acad. Sci. USA* **2006**, *103*, 7829–7834. [CrossRef] [PubMed]
5. Eid, A.; Atlas, J. The role of bisphosphonates in medical oncology and their association with jaw bone necrosis. *Oral. Maxillofac. Surg. Clin. N. Am.* **2014**, *26*, 231–237. [CrossRef] [PubMed]
6. Wasserzug, O.; Kaffe, I.; Lazarovici, T.S.; Weissman, T.; Yahalom, R.; Fliss, D.M.; Yarom, N. Involvement of the maxillary sinus in bisphosphonate-related osteonecrosis of the jaw: Radiologic aspects. *Am. J. Rhinol. Allergy* **2017**, *31*, 36–39. [CrossRef] [PubMed]
7. Zacharis, C.; Tzanavaras, P. Determination of bisphosphonate active pharmaceutical ingredients in pharmaceuticals and biological material: A review of analytical methods. *J. Pharm. Biomed. Anal.* **2008**, *48*, 483–496. [CrossRef] [PubMed]
8. Xie, Z.; Jiang, Y.; Zhang, D.Q. Simple analysis of four bisphosphonates simultaneously by reversed phase liquid chromatography using n-amylamine as volatile ion-pairing agent. *J. Chromatogr. A* **2006**, *1104*, 173–178. [CrossRef] [PubMed]
9. Vallano, P.T.; Shugarts, S.B.; Kline, W.F.; Woolf, E.J.; Matuszewski, B.K. Determination of risedronate in human urine by column-switching ion-pair high-performance liquid chromatography with ultraviolet detection. *J. Chromatogr. B Anal. Technol. Biomed. Life Sci.* **2003**, *794*, 23–33. [CrossRef]
10. Kyriakides, D.; Panderi, I. Development and validation of a reversed-phase ion-pair high-performance liquid chromatographic method for the determination of risedronate in pharmaceutical preparations. *Anal. Chim. Acta* **2007**, *584*, 153–159. [CrossRef] [PubMed]
11. Taylor, G.E. Determination of Impurities in Clodronic Acid by Anion-Exchange Chromatography. *J. Chromatogr. A* **1997**, *770*, 261–271. [CrossRef]
12. Hasan, M.; Schumacher, G.; Seekamp, A.; Taedken, T.; Siegmund, W.; Oswalda, S. LC-MS/MS method for the determination of clodronate in human plasma. *J. Pharm. Biomed. Anal.* **2014**, *100*, 341–347. [CrossRef] [PubMed]
13. Pérez-Ruiz, T.; Martínez-Lozano, C.; García-Martínez, M.D. A sensitive post-column photochemical derivatization/fluorimetric detection system for HPLC determination of bisphosphonates. *J. Chromatogr. A* **2009**, *1216*, 1312–1318. [CrossRef] [PubMed]
14. Lapko, V.N.; Miller, P.S.; Sheldon, C.E.; Nachi, R.; Kafonek, C.J. Quantitative analysis of bisphosphonates in biological samples. *Bioanalysis* **2014**, *6*, 2931–2950. [CrossRef] [PubMed]

15. Bertolini, T.; Vicentini, L.; Boschetti, S.; Andreatta, P.; Gatti, R. A novel automated hydrophilic interaction liquid chromatography method using diode-array detector/electrospray ionization tandem mass spectrometry for analysis of sodium risedronate and related degradation products in pharmaceuticals. *J. Chromatogr. A* **2014**, *1365*, 131–139. [CrossRef] [PubMed]
16. Zhu, L.S.; Lapko, V.N.; Lee, J.W.; Basir, Y.J.; Kafonek, C.; Olsen, R.; Briscoe, C. A general approach for the quantitative analysis of bisphosphonates in human serum and urine by high performance liquid chromatography/tandem mass spectrometry. *Rapid Commun. Mass Spectrom.* **2006**, *20*, 3421–3426. [CrossRef] [PubMed]
17. Raccor, B.S.; Sun, J.; Lawrence, R.F.; Li, L.; Zhang, H.; Somerman, M.J.; Totah, R.A. Quantitation of zoledronic acid in murine bone by liquid chromatography coupled with tandem mass spectrometry. *J. Chromatogr. B Anal. Technol. Biomed. Life Sci.* **2013**, *935*, 54–60. [CrossRef] [PubMed]
18. Wong, A.S.Y.; Ho, E.N.M.; Wan, T.S.M.; Lamb, K.K.H.; Stewart, B.D. Liquid chromatography–mass spectrometry analysis of five bisphosphonates in equine urine and plasma. *J. Chromatogr. B* **2015**, *998–999*, 1–7. [CrossRef]
19. Chen, M.; Liu, K.; Zhong, D.; Chen, X. Trimethylsilyl diazomethane derivatization coupled with solid-phase extraction for the determination of alendronate in human plasma by LC-MS/MS. *Anal. Bioanal. Chem.* **2012**, *402*, 791–798. [CrossRef]
20. Ghassabian, S.; Wright, L.A.; Dejager, A.D.; Smith, M.T. Development and validation of a sensitive solid-phase-extraction (SPE) method using high-performance liquid chromatography/tandem mass spectrometry (LC-MS/MS) for determination of risedronate concentrations in human plasma. *J. Chromatogr. B Anal. Technol. Biomed. Life Sci.* **2012**, *881–882*, 34–41. [CrossRef]
21. Yang, Y.; Liu, C.; Zhang, Y.; Zhou, L.; Zhong, D.; Chen, X. On-cartridge derivatization coupled with solid-phase extraction for the ultra-sensitive determination of minodronic acid in human plasma by LC-MS/MS method. *J. Pharm. Biomed. Anal.* **2015**, *114*, 408–415. [CrossRef] [PubMed]
22. Johnsen, E.; Leknes, S.; Wilson, S.R.; Lundanes, E. Liquid chromatography-mass spectrometry platform for both small neurotransmitters and neuropeptides in blood, with automatic and robust solid phase extraction. *Sci. Rep.* **2015**, *5*, 1–16. [CrossRef] [PubMed]
23. Hemström, P.; Irgum, K. Hydrophilic interaction chromatography. *J. Sep. Sci.* **2006**, *29*, 1784–1821. [CrossRef] [PubMed]
24. Guo, Y. Recent progress in the fundamental understanding of hydrophilic interaction chromatography (HILIC). *Analyst* **2015**, *140*, 6452–6466. [CrossRef] [PubMed]
25. Panderi, I.; Malamos, Y.; Machairas, G.; Zaharaki, S. Investigation of the retention mechanism of cephalosporins by zwitterionic hydrophilic interaction liquid chromatography. *Chromatographia* **2016**, *79*, 995–1002. [CrossRef]
26. Alpert, A.J. Electrostatic repulsion hydrophilic interaction chromatography for isocratic 659 separation of charged solutes and selective isolation of phosphopeptides. *Anal. Chem.* **2008**, *80*, 62–76. [CrossRef] [PubMed]
27. Moravcová, D.; Planeta, J. Monolithic Silica Capillary Columns with Improved Retention and Selectivity for Amino Acids. *Separations* **2018**, *5*, 48. [CrossRef]
28. Barth, H.G. Chromatography Fundamentals, Part III: Retention Parameters of Liquid Chromatography. *LC-GC* **2018**, *36*, 472–473.
29. Kanmatareddy, A.; De Borba, B.; Rohrer, J. *Evaluation of the USP Risedronate Sodium Assay*; Dionex, Application Note 289. Thermo Fisher Scientific: Sunnyvale, CA, USA. Available online: https://assets.thermofisher.com/TFS-Assets/CMD/Application-Notes/AN-289-IC-USP-Risedronate-Sodium-LPN2926-EN.pdf (accessed on September 2016).
30. Sinigaglia, L.; Varenna, M.; Casari, S. Pharmacokinetic profile of bisphosphonates in the treatment of metabolic bone disorders. *Clin. Cases Miner. Bone Metab.* **2007**, *4*, 30–36.
31. Schneider, S. *Analysis of Risedronate According to USP Using the Agilent 1260 Infinity Bio-Inert Quaternary LC System*; Application Note Agilent Technologies, Inc.: Waldbronn, Germany. Available online: https://www.agilent.com/cs/library/applications/5991-2404EN.pdf (accessed on 1 May 2016).

32. Greco, G.; Letzel, T. Main interactions and influences of the chromatographic parameters in HILIC separations. *J. Chromatogr. Sci.* **2013**, *51*, 684–693. [CrossRef]
33. Machairas, G.; Panderi, I.; Geballa-Koukoula, A.; Rozou, S.; Antonopoulos, N.; Charitos, C.; Vonaparti, A. Development and validation of a hydrophilic interaction liquid chromatography method for the quantitation of impurities in fixed-dose combination tablets containing rosuvastatin and metformin. *Talanta* **2018**, *183*, 131–141. [CrossRef] [PubMed]

© 2019 by the authors. Licensee MDPI, Basel, Switzerland. This article is an open access article distributed under the terms and conditions of the Creative Commons Attribution (CC BY) license (http://creativecommons.org/licenses/by/4.0/).

Article

Miniaturized Matrix Solid-Phase Dispersion for the Analysis of Ultraviolet Filters and Other Cosmetic Ingredients in Personal Care Products

Maria Celeiro, Lua Vazquez, J. Pablo Lamas, Marlene Vila, Carmen Garcia-Jares and Maria Llompart *

Laboratory of Research and Development of Analytical Solutions (LIDSA), Department of Analytical Chemistry, Nutrition and Food Science, Faculty of Chemistry, University of Santiago de Compostela, E-15782 Santiago de Compostela, Spain; maria.celeiro.montero@usc.es (M.C.); lua.vazquez.ferreiro@usc.es (L.V.); juanpablo.lamas@usc.es (J.P.L.); marlene.vila@usc.es (M.V.); carmen.garcia.jares@usc.es (C.G.-J.)
* Correspondence: maria.llompart@usc.es; Tel.: +34-881-814-225

Received: 5 April 2019; Accepted: 3 June 2019; Published: 10 June 2019

Abstract: A method based on micro-matrix solid-phase dispersion (µ-MSPD) followed by gas-chromatography tandem mass spectrometry (GC–MS/MS) was developed to analyze UV filters in personal care products. It is the first time that MSPD is employed to extract UV filters from cosmetics samples. This technique provides efficient and low-cost extractions, and allows performing extraction and clean-up in one step, which is one of their main advantages. The amount of sample employed was only 0.1 g and the extraction procedure was performed preparing the sample-sorbent column in a glass Pasteur pipette instead of the classic plastic columns in order to avoid plasticizer contamination. Factors affecting the process such as type of sorbent, and amount and type of elution solvent were studied by a factorial design. The method was validated and extended to other families of cosmetic ingredients such as fragrance allergens, preservatives, plasticizers and synthetic musks, including a total of 78 target analytes. Recovery studies in real sample at several concentration levels were also performed. Finally, the green extraction methodology was applied to the analysis of real cosmetic samples of different nature.

Keywords: UV filters; matrix solid-phase dispersion; µ-MSPD; miniaturized extraction technique; GC–MS/MS; cosmetic analysis; personal care products; fragrance allergens; preservatives; plasticizers; synthetic musks

1. Introduction

The cosmetic industry is one of the fastest growing markets in the world, due to a high demand for cosmetics and personal care products. Manufacturers must innovate to offer attractive and safe products for consumers to stay ahead in a highly competitive sector. Cosmetic formulations usually include a large number or organic compounds, such as fragrances, preservatives, antioxidants, plasticizers, or surfactants among others. One type of these compounds are the ultraviolet filters (UV filters). These substances are intended to protect consumers against the harmful solar radiation and, although their presence is especially important in sunscreens, they can be found in a broad range of daily care products such as creams, hair-care products, lip protectors, make-up, and many others. The widespread inclusion of UV filters in personal care and consumer products increases the human exposure to these compounds. Some of them are considered as endocrine disruptors, with high bioaccumulative properties. In fact, some of them have been recently detected in human breast milk. Nowadays, according to the Annex VI of the Regulation EC No 1223/2009 [1], 26 organic UV filters are allowed for their use in the formulation of cosmetic products, being the maximum concentration permitted in the final product up to 10% (w/w). It is important to note that the Regulation regarding cosmetic products is being continually updated, with the restriction and even prohibition of several compounds each year. Therefore, the cosmetic sector demands the development of reliable, fast and easy to implement analytical methodology to analyze a broad range of cosmetics ingredients. One major drawback for the analysis of cosmetics is sample preparation, since the cosmetic matrices are complex and varied. Besides, the concentration of the different ingredients in cosmetic formulations usually ranges several orders of magnitude, from the ng g^{-1} to thousands of μg g^{-1}.

Most of the reported methodologies for the determination of UV filters in cosmetics deal with the simultaneous analysis of few target compounds. Regarding the sample preparation, solid-liquid or liquid–liquid extraction, or simple dilution, have been the most employed procedures [2–4]. However, since cosmetics are complex mixtures of ingredients, the direct dilution of the samples can negatively affect the chromatographic determination and the chromatographic system, producing damage in the injector, column and detector. Therefore, the use of sample preparation techniques which imply an in-situ clean-up step is a good approach. In this way, matrix solid-phase dispersion (MSPD) has been proposed for the extraction of different families of cosmetic ingredients such as fragrances, preservatives or dyes [5–7].

New trends in sample preparation are focused on the development of miniaturized procedures which complies with the green chemistry principles [8,9], and techniques such as ultrasound-assisted emulsification microextraction (USAEME) or single drop microextraction have been developed [10,11] for the determination of parabens of phthalates. In this way, a miniaturization of the classical MSPD, micro-MSPD (μ-MSPD), employing low-cost material, low amount of sample and organic solvent consumption, has been successfully proposed for the extraction of different compounds such as synthetic musks, preservatives, fragrance allergens, or dyes [12–15] in cosmetics and personal care products. However, to the best of our knowledge MSPD and μ-MSPD have never been applied for the determination of UV filters.

Regarding the analytical determination of UV filters in cosmetic samples, LC-DAD has been the most employed technique [2]. However, the use of other detectors, such as MS, and especially the use of triple quadrupole working under MS/MS provides improved selectivity and sensitivity [16,17].

The main goal of this work is the development of an analytical methodology based on μ-MSPD-GC–MS/MS for the simultaneous determination of 14 multiclass UV filters in cosmetic samples. The main experimental parameters affecting extraction, such as the type of sorbent, and amount and type of extraction solvent have been optimized by means of experimental design. The method was validated and applied to a broad range of cosmetic and personal care products to quantify not only UV filters, but also other families of compounds such as fragrances, preservatives, plasticizers, and synthetic musks, allowing the simultaneous analysis of 78 compounds with very different chemical nature in a single extraction and chromatographic run.

2. Materials and Methods

2.1. Chemicals, Reagents and Materials

The studied UV filters, their Chemical Abstracts Service (CAS) number, retention times, and MS/MS transitions are summarized in Table 1. Target fragrance allergens, preservatives, plasticizers and synthetic musks are shown in Table S1. Ethyl acetate, acetonitrile (ACN) and isooctane were provided by Sigma-Aldrich Chemie GmbH (Steinheim, Germany), methanol (MeOH) was supplied by Scharlab (Barcelona, Spain), and acetone was provided by Fluka Analytical (Steinheim, Germany). Florisil (60–100 μm mesh), and glass wool were purchased from Supelco Analytical (Bellefonte, PA, USA), and sand (200–300 μm mesh) and anhydrous sodium sulphate, Na_2SO_4, (99%) from Panreac (Barcelona, Spain). Individual stock solutions of all the compounds were prepared in acetone, isooctane or methanol. Further dilutions and mixtures were prepared in acetone (spike solutions) or acetonitrile (calibration study). Solutions were stored in amber glass vials at −20 °C. All solvents and reagents were of analytical grade.

Table 1. Studied ultraviolet (UV) filters. CAS number, retention time and mass spectrometry (MS)/MS transitions.

UV Filter	Acronym	CAS	Retention Time (min)	MS/MS Transition (CE [a], eV)		
Ethylhexylsalicylate	EHS	118-60-5	12.85	120.0	→	92.0 (10)
				<u>138.0</u>	→	120.0 (10)
				250.1	→	120.0 (15)
Benzyl salicylate	BS	118-58-1	13.73	91.0	→	39.0 (30)
				91.0	→	65.0 (15)
				<u>228.1</u>	→	91.1 (10)
Homosalate	HMS	118-56-9	13.88	120.0	→	92.0 (10)
				138.0	→	120.0 (10)
				262.2	→	120.0 (15)
Benzophenone-3	BP3	131-57-7	16.22	151.0	→	95.0 (10)
				227.1	→	127.9 (35)
				<u>227.1</u>	→	184.0 (20)
Isoamyl-4-methoxycinnamate	IAMC	71617-10-2	16.38	161.0	→	133.0 (10)
				<u>178.1</u>	→	161.1 (10)
				248.1	→	178.0 (10)
4-methylbenzylidene camphor	4MBC	36861-47-9	16.63	127.9	→	102.0 (20)
				170.6	→	128.1 (15)
				<u>254.1</u>	→	239.2 (10)
Methyl anthranilate	MA	134-20-3	17.66	119.0	→	91.8 (10)
				<u>137.0</u>	→	119.0 (10)
				275.2	→	137.0 (10)
Etocrylene	ETO	5232-99-5	18.22	231.9	→	176.5 (20)
				248.0	→	164.9 (25)
				<u>276.9</u>	→	248.1 (10)
Ethylhexyl-p-aminobenzoic acid	EHPABA	21245-02-3	19.33	148.0	→	104.2 (25)
				165.1	→	148.6 (15)
				<u>277.2</u>	→	164.9 (10)
2-ethylhexyl 4-methoxycinnamate	2EHMC	5466-77-3	19.69	161.0	→	133.1 (10)
				<u>177.9</u>	→	133.1 (20)
				290.2	→	178.1 (10)
Octocrylene	OCR	6197-30-4	21.48	232.0	→	203.0 (20)
				<u>248.0</u>	→	165.0 (30)
				360.2	→	276.1 (20)
Avobenzone	BMDM	70356-09-1	22.44	161.1	→	118.0 (15)
				295.1	→	135.1 (15)
				<u>309.2</u>	→	279.1 (20)
Diethylamino hydroxybenzoyl hexyl benzoate	DHHB	302776-68-7	23.10	382.2	→	280.2 (10)
				382.2	→	298.1 (10
				<u>397.2</u>	⇌	382.2 (10
Drometrizole trisiloxane	DRT	155633-54-8	25.50	221.1	→	73.1 (15)
				<u>369.1</u>	⇌	250.2 (10)
				444.1	→	296.1 (25

[a] CE: collision energy; underlined SRM transitions: quantification transitions.

Metallic, glass materials, dispersing agents (Florisil and sand), Na_2SO_4 and glass wool were maintained at 230 °C for 12 h before use to eliminate possible phthalate contamination. All materials were allowed to cool down, wrapped with aluminum foil, and Florisil, sand, and Na_2SO_4 were kept in desiccator.

2.2. Cosmetic Samples

Cosmetics and personal care products from national and international brands were obtained from local sources. They included sunscreens intended for adults and for children, hair-care products, moisturizing face creams, antiwrinkle creams, make-up, lip protectors, make-up, lipsticks, among others. The samples were kept in their original containers and protected from light at room temperature.

2.3. µ-MSPD Procedure

Cosmetic samples (0.1 g) were exactly weighed into a glass vial. Then, the sample was gently blended with 0.4 g of the drying agent anhydrous Na_2SO_4, and 0.4 g of the corresponding dispersing agent (Florisil or sand), into the vial, using a glass rod, until a homogeneous mixture was obtained (*ca.* 5 min). The mixture was then transferred into a glass Pasteur pipette (approximately 150 mm), with a small amount of glass wool at the bottom, containing 0.1 g of Florisil (to obtain a further degree of fractionation and an in-situ clean-up step). Finally, a small amount of glass wool was placed on top to compress the mixture. Elution with the corresponding solvent (ethyl acetate, ACN, MeOH or the mixture MeOH/acetone (1:1, v/v)) depending on the experiment was made by gravity flow, collecting the extract into a 1 mL or 2 mL volumetric flask. The obtained extracts were diluted 1:10 (v/v) and 1:100 (v/v) in ACN (or even more when necessary), and analyzed by GC–MS/MS. Fortified samples were spiked with 10 µL of the corresponding spiking solution to get the desired final concentration of the target compounds, and submitted to the same process described above. Figure 1 illustrates the described µ-MSPD procedure under the optimal conditions.

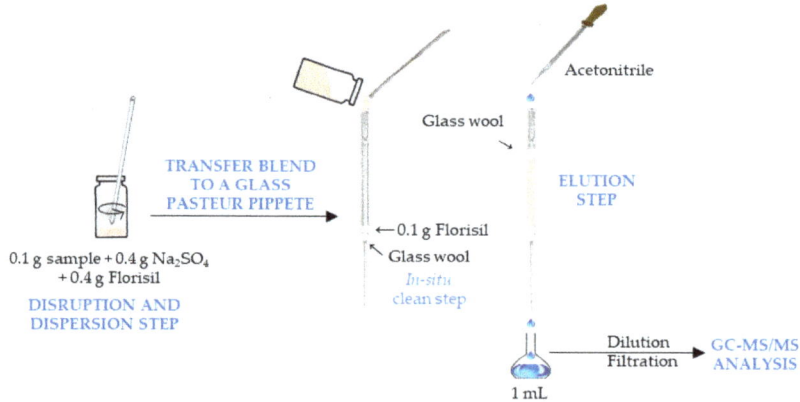

Figure 1. Schematic representation of the micro-matrix solid-phase dispersion (µ-MSPD) procedure under the optimal conditions.

2.4. GC–MS/MS Analysis

The GC–MS/MS analysis was carried out employing a Thermo Scientific Trace 1310 gas chromatograph coupled to a triple quadrupole mass spectrometer (TSQ 8000) with IL 1310 autosampler from Thermo Scientific (San Jose, CA, USA). Separation was performed on a Zebron ZB-Semivolatiles (30 m × 0.25 mm i.d., 0.25 µm film thickness) obtained from Phenomenex (Torrance, CA, USA). Helium (purity 99.999%) was employed as carrier gas at a constant column flow of 1.0 mL min^{-1}. The GC oven temperature was programmed from 60 °C (held 1 min) to 100 °C at 8 °C min^{-1}, to 150 °C at 20 °C min^{-1}, to 200 °C at 25 °C min^{-1} (held 5 min), to 220 °C at 8 °C min^{-1}, and to 290 °C at 30 °C min^{-1} (held 3 min). Pulsed splitless mode (200 kPa, held 1 min) was used for injection and the injector temperature was set at 260 °C. The injection volume was 1 µL and the total run time was 23.5 min.

The mass spectrometer (MSD) was operated in the electron impact (EI) ionization positive mode (+70 eV). The temperatures of the transfer line and the ion source were set at 290 °C and 350 °C, respectively. Selected reaction monitoring (SRM) acquisition mode was implemented monitoring three transitions per compound (see Table 1 for UV filters, and Table S1 for the other compounds). The system was operated by Xcalibur 2.2 and Trace Finder™ 3.2 software.

2.5. Statistical Analysis

Basic and descriptive statistical analysis were performed using Statgraphics Centurion XVII (Manugistics, Rockville, MD, USA) as software package.

3. Results and Discussion

3.1. Chromatographic Analysis

The chromatographic GC–MS/MS method for the determination of the target UV filters was previously proposed by the authors [16–18], and it was extended to other compounds including 25 fragrance allergens, 13 preservatives, 15 plasticizers, and 11 synthetic musks, making a total of 78 compounds. The chromatographic conditions have been previously described in Section 2.4. SRM acquisition mode was employed monitoring two or three transitions per compound (see Table 1 and Table S1).

3.2. Optimization of the μ-MSPD Procedure

The influence of the main parameters potentially affecting the μ-MSPD procedure must be evaluated to obtain an efficient extraction. Several factors, such as the amount of sample, desiccant and dispersing agents were maintained constant, based on previous studies [3,12–14]. The amount of sample was 0.1 g, which was mixed with 0.4 g of Na_2SO_4 to remove the moisture of the samples, which could negatively affect the extraction. Regarding the dispersing agent, its amount was fixed at 0.4 g. The studied parameters were the extraction solvent (factor A), the dispersing agent (factor B), and the extraction volume (factor C), and the different levels are summarized in Table 2. The choice of an appropriate solvent is essential in the development of extraction methods. For an efficient extraction, the solvent must solubilize the target compounds while leaving the sample matrix as intact as possible. Four solvents were investigated: ACN, ethyl acetate (EtAc), methanol (MeOH), and the mixture MeOH/acetone (1:1, v/v). The dispersing agent can be also a very important factor affecting the extraction. In addition, it can contribute to obtain cleaner extracts, preventing lipids and other co-extractable matrix materials from coming out to the extract. Based on our previous works [6,12,14,15], this factor was considered at two levels: Florisil and sand. The solvent volume was also studied at two levels: 1 mL and 2 mL. Larger solvent volumes were not evaluated since the purpose of this study was the development of a green miniaturized extraction protocol. Lower solvent volumes were also not considered since they are not suitable for practical purposes, making necessary the use of inserts to perform further chromatographic analysis.

Table 2. Experimental factors and levels included in the experimental design.

Factor	Code	Level 1	Level 2	Level 3	Level 4
Solvent	A	ACN	EtAc	MeOH	MeOH/acetone (1:1, v/v)
Dispersant	B	Florisil	Sand		
Volume of solvent (mL)	C	1	2		

The influence of the three variables was studied using a multifactor strategy. The study consisted of a multifactor $4*2^2$ design, involving 16 randomized experiments and allowing three degrees of freedom to estimate the experimental error. The design has resolution V, which means that it is capable of evaluating all main effects and all two-factor interactions. Numerical analysis of data resulting from the experimental design was made employing the software package Statgraphics Centurion XVII (Manugistics, CA, USA). The experiments were performed using composite sample prepared as a mixture of four real samples including a sunscreen, a facial cream, a body lotion, and a lip protector. Since the composite sample contained six of the target compounds from the different families of the UV filters studied, it was decided to work with the sample as it, without compounds addition, to really evaluate the capability of the miniaturized procedure to break analyte-matrix interactions, providing efficient extractions. Besides, other compounds such as 11 fragrance allergens, seven preservatives, three plasticizers, and two synthetic musks, were detected in the composite sample. The analysis of variance, ANOVA, describes the impact of the studied factors on the obtained responses. Results for the UV filters are shown in the ANOVA table, Table 3. For the sake of simplicity, only F-ratios and p-values are given. The F-ratio measures the contribution of each factor and interaction on the variance of the response. The p-value tests the statistical significance of each factor and interaction. When p-value is lower than 0.05, the factor has a statistically significant effects at the 95% confidence level.

Table 3. ANOVA summary table obtained for the micro-matrix solid-phase dispersion (µ-MSPD) procedure.

Compound	Solvent (A)		Dispersant (B)		Volume (C)		AB		AC		BC	
	F	p	F	p	F	p	F	p	F	p	F	p
EHS	63	**0.0032**	75	**0.0032**	47	**0.0063**	150	**0.0009**	1.3	0.4114	0.82	0.4313
BP3	157	**0.0008**	8.1	0.0647	49	**0.0059**	422	**0.0002**	4.5	0.1238	0.64	0.4817
IAMC	18	**0.0200**	663	**0.0001**	43	**0.0072**	65	**0.0031**	6.4	0.0802	0.01	0.9361
4MBC	13	**0.0288**	545	**0.0002**	48	**0.0060**	45	**0.0054**	6.3	0.0815	0.75	0.4490
2EHMC	2.6	0.2264	163	**0.0010**	9.7	0.0525	17	**0.0202**	2.2	0.2667	0.03	0.8792
OCR	4.0	0.1425	172	**0.0010**	13	**0.0360**	11	**0.0374**	2.3	0.2560	0.48	0.5392

p-values lower than 0.05 (in bold) denotes statistical significance.

As can be seen, the three studied factors were significant for all the UV filters present in the sample in most cases. The interaction solvent-dispersant (AB) was significant for all the compounds. The other two second order factors (solvent-volume, AC and dispersant-volume, BC) were not significant. Figure 2 shows some selected mean plot graphs, that illustrate the effect of the main factors by showing the mean values as well as the confidence intervals for each level, easily visualizing the most favorable extraction conditions. For all the UV filters, the most efficient solvent was ACN providing higher responses (see Figure 2a). Regarding the dispersing agent, Florisil gave also higher responses for all the analytes (see Figure 2b). As regards the interaction AB, some examples are included in Figure 2c. The two-factor plots display the least squared means at all combinations of two factors, which allows studying the effect of both factors simultaneously. In this case, two different behaviors can be observed. For 2-ethylhexyl 4-methoxycinnamate (2EHMC), 4-methylbenzylidene camphor (4MBC), octocrylene (OCR) and isoamyl-4-methoxycinnamate (IAMC), the use of Florisil provided the highest response regardless of the solvent used (see as example OCR graph in Figure 2c). In the case of ethylhexylsalicylate (EHS) and benzophenone-3 (BP3), the use of sand was more favorable when MeOH or the mixture MeOH/acetone (1:1, v/v) was employed but, in any case, higher responses were obtained using ACN or EtAc with Florisil (see as example BP3 graph in Figure 2c). Regarding solvent volume, 2 mL was initially more favorable, although the differences in the responses were not very high (see Figure 2d).

Since Florisil was the most favorable dispersing agent for all analytes, the results were analyzed considering only the experiments carried out with this sorbent. The ANOVA results were similar for all the analytes and are graphically displayed for IAMC and EHS as example in Figure 3a. The plot shows scaled effects for each factor, so the natural variance of the points in the diagram can be comped to that of the residuals, displayed at the bottom of the plot. By comparing the variability amongst the factors to that of the residuals, it is easy to identify those factors showing differences of a greater magnitude than could be solely accounted by the experimental error. As can be observed, the solvent nature was significant, but the amount of solvent was not a significant factor. The levels of the factors at the right part of the ANOVA plot indicate the conditions that offer higher response and therefore, more efficient extraction. In the mean plot in Figure 3b the influence of the solvent is clearly appreciate. ACN and EtAc provided similar results, whereas for the other solvents the responses were clearly lower.

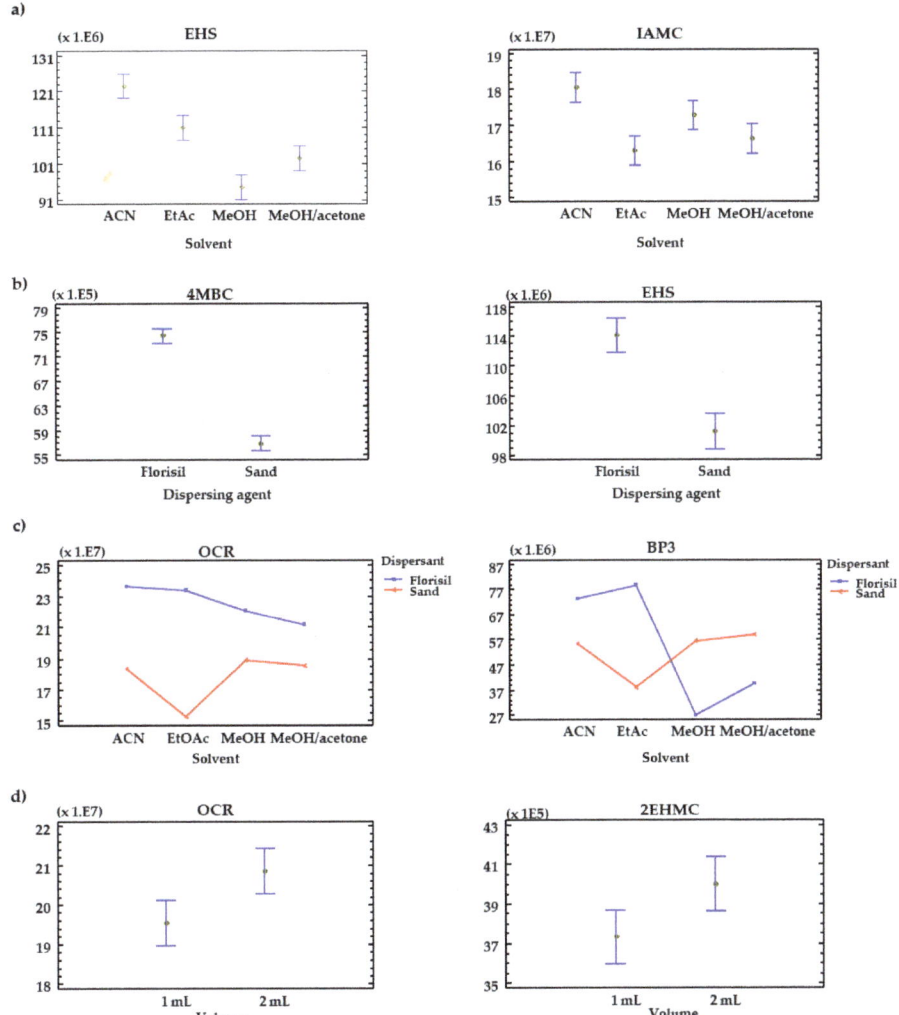

Figure 2. Mean plots (**a,b,d**) and interaction plots (**c**) of the main factors studied in the multi-factor categorical design for some representative ultraviolet (UV) filters.

Therefore, in view of the results, the selected conditions for the analysis of UV filters comprise the use of Florisil as dispersing agent, and ACN or EtAc as eluting solvent. Under these conditions the amount of solvent was not significant and, therefore, the low solvent volume, 1 mL, was selected. Regarding the other cosmetic ingredients and additives present in the composite sample, including fragrance allergens, preservatives, plasticizers and synthetic musks, the statistical analysis showed as more favourable conditions the once previously selected for the UV filters. Therefore, a general multianalyte method for the determination of all these families of personal care products (PCPs) can be proposed.

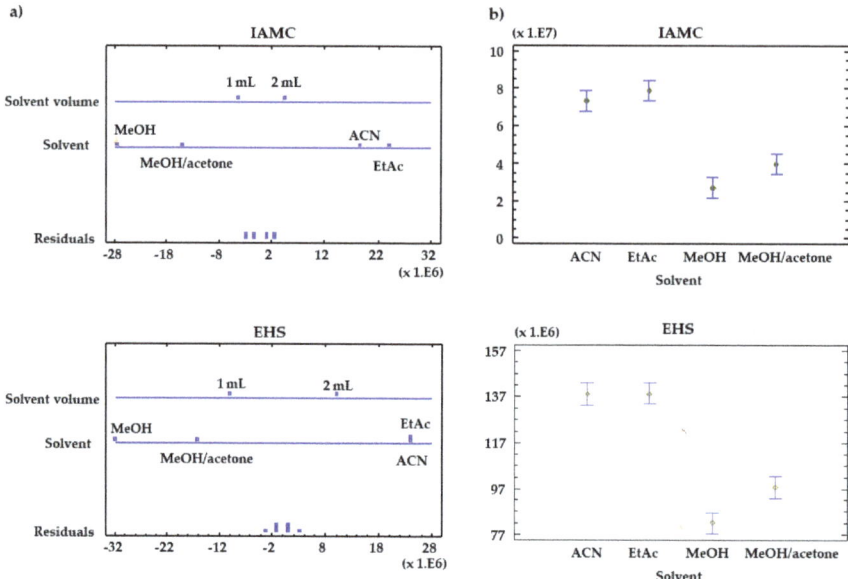

Figure 3. (**a**,**b**) ANOVA plots showing the main effects for isoamyl-4-methoxycinnamate (IAMC) and ethylhexylsalicylate (EHS).

3.3. Method Performance

The µ-MSPD-GC–MS/MS method was validated in terms of linearity, accuracy and precision. Method performance is summarized in Table 4 for UV filters, and Table 5 for the other compounds.

Table 4. µ-MSPD-gas-chromatography tandem mass spectrometry (GC–MS/MS) performance for the UV filters. Linearity, precision, and recovery studies.

UV Filters	Linearity		Precision [a]		Recoveries		
	Range (mg L^{-1})	R^2	RSD, %	Mean Values	100 µg g^{-1}	10 µg g^{-1}	1 µg g^{-1}
EHS	0.001–10	0.9999	10	109 ± 11	106 ± 2	111 ± 4	110 ± 5
BS	0.002–10	0.9999	4.8	110 ± 3	111 ± 2	116 ± 6	103 ± 1
HMS	0.002–10	0.9999	10	109 ± 6	110 ± 2	109 ± 7	108 ± 10
BP3	0.002–10	0.9980	3.9	106 ± 6	103 ± 3	117 ± 10	98.7 ± 5.3
IAMC	0.001–10	0.9992	14	98.4 ± 5.8	100 ± 2	102 ± 6	93.3 ± 9.5
4MBC	0.002–10	0.9997	8.0	97.9 ± 6.7	97.9 ± 2.8	99.4 ± 7.2	96.6 ± 10.0
MA	0.001–10	0.9994	5.8	106 ± 5	104 ± 2	99.4 ± 8.2	114 ± 4
ETO	0.001–10	0.9998	5.2	97.9 ± 7.3	97.4 ± 7.2	93.3 ± 9.7	103 ± 5
EHPABA	0.002–10	0.9997	8.6	99.0 ± 4.3	101 ± 2	95.2 ± 6.8	101 ± 4
2EHMC	0.002–10	0.9992	10	99.5 ± 4.1	99.4 ± 1.5	99.1 ± 8.5	100 ± 3
OCR	0.002–10	0.9999	9.8	104 ± 4	104 ± 4	n.c. [b]	n.c. [b]
BMDM	1–1000	0.9966	6.1	111 ± 2	111 ± 2	n.c. [c]	n.c. [c]
DHHB	1–50	0.9922	10	108 ± 3	108 ± 3	n.c. [c]	n.c. [c]
DRT	0.1–100	0.9915	5.6	98.7 ± 2.3	98.2 ± 1.2	97.4 ± 3.5	n.c [c]

[a] n = 6; [b] not calculated since the compound was detected in the sample or [c] below linear range.

The calibration study was performed employing standard solutions prepared in acetonitrile containing the 78 compounds at different levels, covering a concentration range from 0.001 to 10 mg L^{-1} (see specific ranges for each compound in Tables 4 and 5) with twelve levels and three replicates per level. The method exhibited a direct proportional relationship between the concentration of each analyte and the chromatographic response with determination coefficients $R^2 \geq 0.9915$ for all compounds. Calibration plots for some representative compounds are shown in Figure S1.

Table 5. μ-MSPD-GC–MS/MS performance for the fragrance allergens, preservatives, plasticizers and synthetic musks. Linearity, precision, and recovery studies.

Compounds	Linearity Range (mg L^{-1})	R^2	Precision [a] RSD, %	Mean Values	Recoveries 100 μg g^{-1}	10 μg g^{-1}
Fragrance allergens						
Pinene	0.001–10	0.9994	3.5	70.2 ± 6.0	77.8 ± 7.8	62.6 ± 4.2
Limonene	0.001–10	0.9985	7.2	85.1 ± 3.7	97.1 ± 3.6	73.1 ± 3.8
Benzyl alcohol	0.001–10	0.9982	9.7	109 ± 6	107 ± 2	111 ± 9
Linalool	0.005–10	0.9994	6.7	98.6 ± 6.7	104 ± 2	93. 2 ± 11.4
Methyl-2-octynoate	0.1–10	0.9999	6.0	106 ± 5	105 ± 2	107 ± 8
Citronellol	0.05–10	0.9999	8.8	107 ± 6	107 ± 2	107 ± 10
Citral	0.002–10	0.9994	7.1	99.5 ± 4	112 ± 1	86.9 ± 7.0
Geraniol	0.02–10	0.9998	6.9	106 ± 3	96.4 ± 0.4	116 ± 6
Cinnamaldehyde	0.005–10	0.9999	6.8	101 ± 7	106 ± 2	95.7 ± 11.3
Hydroxycitronellal	0.005–10	0.9995	6.0	108 ± 2	100 ± 2	116 ± 3
Anise alcohol	0.01–10	0.9998	8.3	102 ± 5	101 ± 2	103 ± 9
Cinnamyl alcohol	0.001–10	0.9996	8.7	105 ± 6	105 ± 4	105 ± 7
Eugenol	0.005–10	0.9965	6.4	108 ± 4	105 ± 3	111 ± 5
Methyleugenol	0.005–10	0.9981	6.8	95.6 ± 3.7	102 ± 2	89.2 ± 5.4
Isoeugenol	0.02–10	0.9992	8.4	100 ± 4	103 ± 3	97.0 ± 5.3
Coumarin	0.02–10	0.9980	6.6	102 ± 8	104 ± 3	100 ± 13
α-isomethylionone	0.005–10	0.9975	8.6	99.5 ± 5.6	101 ± 2	98.1 ± 9.3
Lilial®	0.005–10	0.9995	6.6	100 ± 7	106 ± 2	94.2 ± 12.1
Amylcinnamaldehyde	0.005–10	0.9991	8.1	106 ± 4	106 ± 1	106 ± 7
Lyral®	0.002–2	0.9971	7.2	107 ± 4	108 ± 1	105 ± 7
Amylcinnamyl alcohol	0.005–10	0.9992	8.1	110 ± 7	107 ± 1	112 ± 12
Farnesol	0.02–10	0.9994	12	107 ± 7	106 ± 4	107 ± 10
Hexylcinnamaldehyde	0.01–10	0.9922	8.1	107 ± 5	107 ± 2	106 ± 7
Benzyl benzoate	0.002–10	0.9992	6.7	102 ± 6	104 ± 2	99.3 ± 10.2
Benzyl cinnamate	0.001–10	0.9999	5.4	103 ± 4	103 ± 3	102 ± 5
Preservatives						
Bronidox	0.002–10	0.9999	4.8	103 ± 6	110 ± 1	95.6 ± 11.1
Phenoxyethanol (PhEtOH)	0.001–10	0.9999	7.4	110 ± 7	101 ± 1	120 ± 14
Methyl paraben (MeP)	0.001–10	0.9997	10	110 ± 6	102 ± 2	117 ± 11
Butylhydroxyanisole (BHA)	0.0001–10	0.9990	5.5	95.6 ± 4.5	103 ± 3	88.1 ± 5.9
Butylhydroxytoluene (BHT)	0.005–10	0.9996	4.9	95.1 ± 3.3	103 ± 2	87.2 ± 4.7
Ethyl paraben (EtP)	0.02–10	0.9999	11	102 ± 8	100 ± 2	103 ± 14
Isopropyl paraben (iPrP)	0.05–10	0.9995	10	103 ± 4	105 ± 3	99.9 ± 4.6
Propyl paraben (PrP)	0.01–10	0.9998	11	98.5 ± 4.2	102 ± 2	94.9 ± 6.4
Iodopropynylbutyl carbamate (IPBC)	0.002–10	0.9997	3.5	104 ± 9	103 ± 2	105 ± 15
Isobutyl paraben (iBuP)	0.005–10	0.9999	8.1	103 ± 2	102 ± 2	104 ± 2
Butyl paraben (BuP)	0.005–10	0.9999	7.6	99.6 ± 1.9	101 ± 1	98.3 ± 2.8
Triclosan (TCS)	0.002–10	0.9983	2.3	115 ± 9	113 ± 3	117 ± 14
Benzyl paraben (BzP)	0.05–10	0.9995	4.2	101 ± 6	99.2 ± 3.3	103 ± 8
Plasticizers						
Dimethyl adipate (DMA)	0.01–10	0.9998	5.3	116 ± 5	104 ± 1	118 ± 8
Diethyl adipate (DEA)	0.001–10	0.9989	6.5	98.5 ± 7.7	103 ± 2	93.9 ± 13.4
Diethyl phthalate (DEP)	0.005–10	0.9984	6.2	99.9 ± 4.0	102 ± 1	97.8 ± 7.1
Diisobutyl phthalate (DIBP)	0.001–10	0.9992	4.7	98.9 ± 4.9	101 ± 2	96.8 ± 7.8
Dibutyl phthalte (DBP)	0.001–10	0.9997	9.1	100 ± 5	102 ± 2	98.4 ± 8.7
Dimethoxyethyl phthalate (DMEP)	0.005–10	0.9999	6.3	105 ± 6	105 ± 2	105 ± 9
Diisopentyl phthalate (DIPP)	0.002–10	0.9998	8.2	99.3 ± 5.5	100 ± 3	98.7 ± 7.9
Dipentyl phthalate (DPP)	0.001–10	0.9997	11	101 ± 3	101 ± 2	101 ± 3
Benzylbutyl phthalate (BBP)	0.002–10	0.9997	13	99.7 ± 5.5	100 ± 3	99.5 ± 8.0
Diethylhexyl adipate (DEHA)	0.005–10	0.9997	14	95.4 ± 5.4	96.2 ± 2.8	94.6 ± 8.0
Diisoheptyl phthalate (DIHP)	0.002–10	0.9998	3.9	97.6 ± 5.3	100 ± 3	95.3 ± 7.6
Dicyclohexyl phthalate (DCHP)	0.005-10	0.9996	9.5	101 ± 4	101 ± 3	101 ± 4
Diethylhexyl phthalate (DEHP)	0.01–10	0.9997	9.2	100 ± 3	100 ± 4	100 ± 1
Diphenyl phthalate (DPhP)	0.001–10	0.9999	11	101 ± 5	98.7 ± 4.5	103 ± 5
Di-n-octyl phthalate (DnOP)	0.005–10	0.9998	8.0	105 ± 3	107 ± 5	102 ± 1
Synthetic musks						
Cashmeran	0.001–10	0.9976	7.1	100 ± 4	103 ± 2	97.0 ± 5.0
Celestolide	0.002–10	0.9979	5.4	100 ± 8	106 ± 2	94.2 ± 13.8
Phantolide	0.005–10	0.9977	5.9	99.8 ± 7.3	104 ± 2	95.6 ± 12.6
Ambrette	0.005–10	0.9998	10	97.7 ± 7	104 ± 1	91.4 ± 13.0
Trasolide	0.1–10	0.9996	9.6	102 ± 9	98.3 ± 9.9	105 ± 9
Galaxolide	0.001–10	0.9995	6.7	97.3 ± 5.1	101 ± 3	97.3 ± 8.2
Tonalide	0.005–10	0.9988	7.2	96.1 ± 7.2	101 ± 1	91.3 ± 13.5
Musk Moskene	0.002–10	0.9999	9.4	108 ± 5	112 ± 2	103 ± 8
Musk Tibetene	0.005–10	0.9995	7.1	99.1 ± 4.9	104 ± 1	94.3 ± 8.8
Ambrettolide	0.002–10	0.9988	7.0	105 ± 3	104 ± 3	106 ± 2
Musk Ketone	0.001–10	0.9998	10	98.4 ± 7.1	90.8 ± 3.4	98.0 ± 10.7

[a] n = 6.

Intra-day, and inter-day precision was also evaluated. The relative standard deviation (RSD) values for the inter-day are shown in Tables 4 and 5, and they were lower than 10% for all the analyzed UV filters, and lower than 14% for the other compounds.

Recovery studies were carried out by implementing the optimized μ-MSPD-GC–MS/MS method to a real cosmetic sample (a moisturizing hand cream). Sample was fortified at three different concentration levels (1, 10 and 100 µg g^{-1}) for the UV filters and the μ-MSPD-GC–MS/MS procedure was performed. Recoveries were calculated as the ratio of concentration found/added considering the responses obtained for each analyte, and they are shown in Table 4 Quantitative recoveries were obtained in all cases, with mean values between 97% and 111%. The precision was also evaluated, and the obtained relative standard deviation (RSD) values were lower than 10% for all the analytes.

Recovery studies were also performed for the fragrance allergens, preservatives, plasticizers and synthetic musks, at two different concentration levels (10 and 100 µg g^{-1}). Results are summarized in Table 5. As can be seen, good recoveries with mean values between 70% and 110% were obtained for all the studied compounds. The RSD values were also lower than 10% in all cases.

3.4. Application to Real Samples

To show the suitability of the proposed methodology, 13 different cosmetic and personal care products were analyzed, including moisturizing face creams, sunscreens with different solar protection factor (SPF), including products intended from children, blemish base (BB) creams, hair-care products, protection lipsticks, hands cream, make-up, or vitalizing creams. Concentration (µg g^{-1}) of the target UV filters, and the other analyzed PCPs are summarized in Table 6.

Eleven out of the 14 studied UV filters were detected in the analyzed samples. The UV filter most frequently found was 2-EHMC, in 11 of the 13 samples, with concentration levels up to 46,364 µg g^{-1} (4.6%, w/w) followed by EHS in eight samples. The concentration for this UV filter was higher than 20,000 µg g^{-1} (2%, w/w) in four samples (S2, S4, S5, and S7). OCR and avobenzone (BMDM) were found in seven samples, at concentrations up to 50,000 µg g^{-1}, excluding BMDM in samples S2, and S4. The other UV filters homosalate (HMS), BP3, benzyl salicylate (BS), and IAMC, were found in six, five, four and three samples, respectively, with concentration ranging from 0.5 to 52,000 µg g^{-1}, whereas 4MBC, DHHB and DRT were only found in one sample each one. Regarding the number of compounds per sample, sample S3 (BB cream) contained eight out of the 11 detected UV filters, followed by sample S2, sample S6 and sample S8, which contained 6 compounds. Highlights especially the high UV filters concentration (between 25,000–99,000 µg g^{-1}) found in Sample S2. This sample was a SPF 50 sunscreen. In the other samples, between 1–5 UV filters were detected. Although for some compounds, while very high concentrations were found, all of them comply with the European requirements according to the Regulation EC No 1223/2009 [1].

Regarding the other studied PCPs, 14 of the 25 target fragrance allergens were found. Highlights the presence of limonene and benzyl alcohol in 12 of the 13 analyzed samples, with concentrations ranging between 0.2 to 213 µg g^{-1}. The other fragrance allergens were found in between 1–4 samples. It is important to note the presence of Lyral®, fragrance which has been recently banned, in one cream at 87 µg g^{-1}. Sample S3, a BB cream, contained the highest number of fragrances, nine of them at also the highest concentration for them, 270 µg g^{-1} for α-isomethylionone. The other analyzed samples contained between one (sample S2) and six (samples S4 and S9) fragrances.

Table 6. Concentration of the UV filters and the other personal care products (PCPs) ($\mu g\,g^{-1}$ equivalent to $\times 10^4$ %w/w).

	S1	S2	S3	S4	S5	S6	S7	S8	S9	S10	S11	S12	S13
UV filters													
EHS		26923 ± 2851	6 ± 1	39706 ± 1131	28372 ± 698	17 ± 3	23925 ± 3115	12 ± 1	1.1 ± 0.1				
BS	8.8 ± 0.3		17 ± 2			1.7 ± 0.5	0.5 ± 0.1						
HMS	0.5 ± 0.1		1.2 ± 0.3	52597 ± 2980		1.4 ± 0.2		8.4 ± 0.4	6.5 ± 0.1				
BP3			1.0 ± 0.2			46 ± 1			3 ± 1		18 ± 1		4693 ± 1727
IAMC			1.8 ± 0.1			6 ± 2			24 ± 1				
4MBC		27061 ± 3013											
2-EHMC	4927 ± 272	46364 ± 3939	350 ± 77	12 ± 3		17230 ± 3233		158 ± 4	46154 ± 3290	3 ± 1	0.9 ± 0.07	4 ± 1	1 ± 0.07
OCR		49327 ± 4146	7722 ± 1063	28 ± 10	29378 ± 1118		14065 ± 2442	42633 ± 2059				3 ± 0.1	
BMDM	2970 ± 116	66444 ± 20047	3260 ± 763	86318 ± 35293	53437 ± 4486		19397 ± 7542	19490 ± 3001					
DHHB		99111 ± 17536											
DRT								13300 ± 820					
Fragrance allergens	S1	S2	S3	S4	S5	S6	S7	S8	S9	S10	S11	S12	S13
Limonene	61 ± 5		2.1 ± 0.4	281 ± 35	0.4 ± 0.01	17 ± 1	0.3 ± 0.04	0.6 ± 0.01	4.3 ± 0.3	0.5 ± 0.02	18 ± 2	0.3 ± 0.01	2132 ± 120
Benzyl alcohol	3.6 ± 0.2		0.7 ± 0.1	2.5 ± 0.4	4.9 ± 0.1	1.8 ± 0.5	1.2 ± 0.1	0.7 ± 0.04	1.2 ± 0.1	0.4 ± 0.01	0.2 ± 0.01	0.3 ± 0.01	113 ± 40
Linalool	120 ± 7		4.6 ± 0.6	234 ± 22				0.7 ± 0.01	2.0 ± 0.1				127 ± 50
Citronellol				34 ± 4									
Citral							12 ± 2						34 ± 11
Hydroxycitronellal			2.0 ± 0.2						31 ± 2				
Cinnamyl alcohol	12 ± 1		1.1 ± 0.1		0.9 ± 0.02								
Eugenol	5.7 ± 0.5			22 ± 3									
Coumarin	4.9 ± 0.4		270 ± 32	55 ± 7									
α-isomethylionone	6.6 ± 0.4				80 ± 1				1.1 ± 0.3				
Lilial®			87 ± 12										
Lyral®	63 ± 4		0.8 ± 0.1		3.9 ± 0.2	1.7 ± 0.5			5.5 ± 0.6		20 ± 2	6.7 ± 0.02	
Farnesol													
Hexylcinnamaldehyde	2.2 ± 0.2	1.0 ± 0.1	11 ± 1			0.7 ± 0.1	0.8 ± 0.1			1.1 ± 0.1			
Benzyl benzoate													
Preservatives	S1	S2	S3	S4	S5	S6	S7	S8	S9	S10	S11	S12	S13
PhEtOH	d	8461 ± 1164	2384 ± 275		6029 ± 178	6181 ± 1673	3663 ± 526	47 ± 1	88 ± 3	6.1 ± 0.2	6660 ± 1323	1608 ± 52	3650 ± 153
MeP	3094± 244	0.4 ± 0.1	5.4 ± 0.1			2778 ± 615	1.4 ± 0.4		1382 ± 46			0.3 ± 0.001	978 ± 356
BHA													
BHT	3.1 ± 0.4	52 ± 7	31 ± 3	1.0 ± 0.2	0.9 ± 0.1	2.0 ± 0.5	1.2 ± 0.1	0.9 ± 0.0004	20 ± 1		1 ± 0.2	69 ± 2	80 ± 31
EtP	895 ± 73					644 ± 131			6.9 ± 0.1				226 ± 81
PrP	793 ± 57					318 ± 71			545 ± 18				100 ± 36
iBuP						436 ± 13							110 ± 31
BuP	947 ± 54					763 ± 199			3.9 ± 0.7				209 ± 81

Table 6. *Cont.*

Plasticizers	S1	S2	S3	S4	S5	S6	S7	S8	S9	S10	S11	S12	S13
DEP		13 ± 2	396 ± 43			3.4 ± 1.0			26 ± 1	0.7 ± 0.1			
DBP			1.7 ± 0.2			3.8 ± 0.9			15 ± 1				
DEHA	3.2 ± 0.1	52 ± 4	26305 ± 2379	2.6 ± 0.1	2.6 ± 0.2	45 ± 22	2.4 ± 0.2	2.9 ± 0.04	24 ± 1				
DEHP	9 ± 3	6.8 ± 0.4	9 ± 1	5.0 ± 0.3	5 ± 2	54 ± 16	2.8 ± 0.5	5.7 ± 0.2	51 ± 2				
Synthetic musks	**S1**	**S2**	**S3**	**S4**	**S5**	**S6**	**S7**	**S8**	**S9**	**S10**	**S11**	**S12**	**S13**
Celestolide													
Galaxolide			534 ± 57			1.8 ± 0.2			27 ± 1				
Ambrettolide							12.6 ± 0.3		2.0 ± 0.04				

S1: moisturizing facial cream; S2: SPF 50 sunscreen; S3: BB cream; S4: SPF 50 sunscreen intended for children; S5: leave-on hair serum; S6: moisturizing make-up; S7: anti-wrinkle facial cream; S8: solar stick; S9: antiaging hand and nail cream: S10: lipstick; S11: facial cream; S12: make-up; S13: vitalizing cream.

Seven of the 13 target preservatives were found in the analyzed samples. The most frequently found were phenoxyethanol (PhEtOH) and butylhydroxytoluene (BHT) in 92% of the analyzed samples. The highest PhEtOH concentration reached up to 8461 $\mu g\ g^{-1}$, close to its legal limit (10,000 $\mu g\ g^{-1}$), in sample S2, whereas for BHT its concentration was lower than 80 $\mu g\ g^{-1}$ in all cases. Methyl paraben (MeP) was found in nine samples, reaching 3100 $\mu g\ g^{-1}$, also close to its maximum permitted concentration (4000 $\mu g\ g^{-1}$), in sample S1, whereas the other parabens (EtP, PrP, BuP, and iBuP) were found in six, five, and three samples respectively. The samples containing more preservatives were sample S6 and sample S7, containing both seven preservatives, whereas on the other hand, samples S4 and S10 only contained BHT and PhEtOH, respectively.

Regarding the synthetic musks, only celestolide, cashmeran and ambrettolide were detected in the analyzed samples. Galaxolide was found in three samples at concentrations up to 534 $\mu g\ g^{-1}$, whereas the other two were only detected in one sample each one.

Only four plasticizers out of the 15 studied were detected in the analyzed samples. The diethylhexyl adipate (DEHA) was found in nine samples, with concentrations up to 2630 $\mu g\ g^{-1}$. Regarding the other detected phthalates, DEP was found in five samples, whereas two of the phthalates forbidden for their use as ingredients in cosmetics according to the Regulation EC No 1223/2009, dibutyl phthalate (DBP) and diethylhexyl phthalate (DEHP) were found in three and nine samples, respectively. The detected concentrations were lower than 9 $\mu g\ g^{-1}$ in all samples, and the presence of these compounds may be related with a possible transfer between the plastic package and the cosmetic.

4. Conclusions

A new analytical methodology based on μ-MSPD-GC–MS/MS has been proposed for the first time for the simultaneous analysis of 14 multiclass organic UV filters in cosmetic and personal care products. The main parameters affecting μ-MSPD extraction have been optimized to obtain the highest extraction efficiency. Under the optimal conditions, which implies the use of Florisil as the dispersing agent and 1 mL of ACN as elution solvent, the method was successfully validated in terms of linearity, accuracy and precision. The proposed methodology was extended to other PCPs families, including fragrance allergens, preservatives, plasticizers and synthetic musks comprising a total of 78 compounds. Finally, to show the method suitability, it was applied to a broad range of real cosmetic samples present on the market, including sunscreen, make up, and hair-care products, among many others. In summary, the developed methodology provides a suitable, green, and fast tool to determine a broad range of cosmetic ingredients in a wide variety of cosmetic products, allowing simultaneous analysis of 78 compounds with very different chemical nature in a single extraction and chromatographic run.

Supplementary Materials: The following are available online at http://www.mdpi.com/2297-8739/6/2/30/s1, Figure S1: Calibration plots for some representative compounds of each studied family, Table S1: Retention time and MS/MS transitions for the fragrance allergens, preservatives, plasticizers and synthetic musks.

Author Contributions: Conceptualization, M.L. and M.C.; methodology, M.L. and C.G.J.; validation, J.P.L., L.V. and M.L.; formal analysis, M.V., J.P.L., M.C. and L.V.; resources, M.L. and C.G.J.; writing—original draft preparation, M.L. and M.C.; writing—review and editing, M.C. and M.L.; supervision, M.L.; funding acquisition, M.L.

Funding: This research was supported by the project UNST10-1E-491 (Ministry of Science and Innovation, Spain). The authors belong to the Galician Competitive Research Group GPC2017/04 and to the CRETUS Strategic Partnership (ED431 2018/01). All these programs are co-funded by FEDER (EU).

Conflicts of Interest: The authors declare no conflict of interest.

References

1. European Union. Regulation (EC) No 1223/2009 of the European Parliament and of the Council of 30 November 2009 on cosmetic products. *Off. J. Eur. Union.* **2009**, *342*, 59–209. Available online: https://eur-lex.europa.eu/legal-content/EN/ALL/?uri=CELEX%3A32009R1223 (accessed on 19 April 2019).
2. Salvador, A.; Chisvert, A. Sunscreen analysis: A critical survey on UV filters determination. *Anal. Chim. Acta* **2005**, *537*, 1–14. [CrossRef]

3. Lores, M.; Llompart, M.; Alvarez-Rivera, G.; Guerra, E.; Vila, M.; Celeiro, M.; Lamas, J.P.; Garcia-Jares, C. Positive lists of cosmetic ingredients: Analytical methodology for regulatory and safety controls-A. *Anal. Chim. Acta* **2016**, *915*, 1–26. [CrossRef] [PubMed]
4. Zhong, Z.; Li, G. Current trends in sample preparation for cosmetic analysis. *J. Sep. Sci.* **2017**, *40*, 152–169. [CrossRef] [PubMed]
5. Alvarez-Rivera, G.; Dagnac, T.; Lores, M.; Garcia-Jares, C.; Sanchez-Prado, L.; Lamas, J.P.; Llompart, M. Determination of isothiazolinone preservatives in cosmetics and household products by matrix solid-phase dispersion followed by high-performance liquid chromatography-tandem mass spectrometry. *J. Chromatogr. A* **2012**, *127*, 41–50. [CrossRef] [PubMed]
6. Sanchez-Prado, L.; Lamas, J.P.; Alvarez-Rivera, G.; Lores, M.; Garcia-Jares, C.; Llompart, M. Determination of suspected fragrance allergens in cosmetics by matrix solid-phase dispersion gas chromatography-mass spectrometry analysis. *J. Chromatogr. A* **2011**, *1218*, 5055–5062. [CrossRef] [PubMed]
7. Chen, M.; Bai, H.; Zhai, J.; Meng, X.; Guo, X.; Wang, C.; Wang, P.; Lei, H.; Niu, Z.; Ma, Q. Comprehensive screening of 63 coloring agents in cosmetics using matrix solid-phase dispersion and ultra-high-performance liquid chromatography coupled with quadrupole-Orbitrap high-resolution mass spectrometry. *J. Chromatogr. A* **2019**, *1590*, 27–38. [CrossRef] [PubMed]
8. Anastas, P.; Eghbali, N. Green chemistry: Principles and practice. *Chem. Soc. Rev.* **2010**, *39*, 301–312. [CrossRef] [PubMed]
9. Mohamed, H.M. Green, environment-friendly, analytical tools give insights in pharmaceuticals and cosmetics analysis. *TrAC-Trend. Anal. Chem.* **2015**, *66*, 176–192. [CrossRef]
10. Kamarei, F.; Ebrahimzadeh, H.; Yamini, Y. Optimization of ultrasound-assisted emulsification microextraction with solidification of floating organic droplet followed by high performance liquid chromatography for the analysis of phthalate esters in cosmetic and environmental water samples. *Microchem. J.* **2011**, *99*, 26–33. [CrossRef]
11. Saraji, M.; Mirmahdieh, S. Single-drop microextraction followed by in-syringe derivatization and GC-MS detection for the determination of parabens in water and cosmetic products. *J. Sep. Sci.* **2009**, *32*, 988–995. [CrossRef] [PubMed]
12. Celeiro, M.; Guerra, E.; Lamas, J.P.; Lores, M.; Garcia-Jares, C.; Llompart, M. Development of a multianalyte method based on micro-matrix-solid-phase dispersion for the analysis of fragrance allergens and preservatives in personal care products. *J. Chromatogr. A* **2014**, *1344*, 1–14. [CrossRef] [PubMed]
13. Celeiro, M.; Lamas, J.; Llompart, M.; Garcia-Jares, C. In-vial micro-matrix-solid phase dispersion for the analysis of fragrance allergens, preservatives, plasticizers, and musks in cosmetics. *Cosmetics* **2014**, *1*, 171–201. [CrossRef]
14. Guerra, E.; Celeiro, M.; Lamas, J.P.; Llompart, M.; Garcia-Jares, C. Determination of dyes in cosmetic products by micro-matrix solid phase dispersion and liquid chromatography coupled to tandem mass spectrometry. *J. Chromatogr. A* **2015**, *1415*, 27–37. [CrossRef] [PubMed]
15. Llompart, M.; Celeiro, M.; Lamas, J.P.; Sanchez-Prado, L.; Lores, M.; Garcia-Jares, C. Analysis of plasticizers and synthetic musks in cosmetic and personal care products by matrix solid-phase dispersion gas chromatography-mass spectrometry. *J. Chromatogr. A* **2013**, *1293*, 10–19. [CrossRef] [PubMed]
16. Vila, M.; Celeiro, M.; Lamas, J.P.; Dagnac, T.; Llompart, M.; Garcia-Jares, C. Determination of fourteen UV filters in bathing water by headspace solid-phase microextraction and gas chromatography-tandem mass spectrometry. *Anal. Methods* **2016**, *8*, 7069–7079. [CrossRef]
17. Vila, M.; Lamas, J.P.; Garcia-Jares, C.; Dagnac, T.; Llompart, M. Optimization of an analytical methodology for the simultaneous determination of different classes of ultraviolet filters in cosmetics by pressurized liquid extraction-gas chromatography tandem mass spectrometry. *J. Chromatogr. A* **2015**, *1405*, 12–22. [CrossRef] [PubMed]
18. Vila, M.; Lamas, J.P.; Garcia-Jares, C.; Dagnac, T.; Llompart, M. Ultrasound-assisted emulsification microextraction followed by gas chromatography-mass spectrometry and gas chromatography-tandem mass spectrometry for the analysis of UV filters in water. *Microchem. J.* **2016**, *124*, 530–539. [CrossRef]

© 2019 by the authors. Licensee MDPI, Basel, Switzerland. This article is an open access article distributed under the terms and conditions of the Creative Commons Attribution (CC BY) license (http://creativecommons.org/licenses/by/4.0/).

Article

Treatment of Tannery Wastewater with Vibratory Shear-Enhanced Processing Membrane Filtration

Anastasios I. Zouboulis *, Efrosyni N. Peleka and Anastasia Ntolia

Laboratory of Chemical and Environmental Chemical Technology, Department of Chemistry,
Aristotle University, GR-54124 Thessaloniki, Greece; peleka@chem.auth.gr (E.N.P.); antolia@cheng.auth.gr (A.N.)
* Correspondence: zoubouli@chem.auth.gr

Received: 8 January 2019; Accepted: 19 March 2019; Published: 8 April 2019

Abstract: The performance of a vibratory shear-enhanced process (VSEP) combined with an appropriate membrane unit for the treatment of simulated or industrial tannery wastewaters was investigated. The fundamental operational and pollution parameters were evaluated, i.e., the membrane type, the applied vibration amplitude, as well as the removal rates (%) of tannins, chemical oxygen demand (COD), N_{total}, turbidity and color. Regarding the system's treatment efficiency, specific emphasis was given towards the removal of organics (expressed as COD values), suspended solids (SS), conductivity (as an index of dissolved solids' presence) and total nitrogen. The removal of organic matter in terms of COD exceeded 75% for all the examined cases. The quality of treated wastewater was affected not only by the membrane specific type (i.e., the respective pore diameters), but also by the applied vibration amplitude. Furthermore, an average 50% removal rate, regarding the aforementioned parameters, was observed both for the simulated and the industrial tannery wastewaters during the microfiltration (MF) experiments. That removal rate was further increased up to 85%, when ultrafiltration (UF) was applied, and up to 99% during the Reverse Osmosis (RO) experiments, considering the maximum applied vibration amplitude (31.75 mm).

Keywords: membrane filtration-treatment; membrane type-operation; membrane fouling mechanism; tannery industrial wastewater; vibratory shear-enhanced process (VSEP)

---

1. Introduction

The leather tanning industry is a globalized industry and the European Union (EU) tanners are highly dependent on access to raw materials and export markets. The EU tanning industry is still the world's largest leather supplier in the international market. This is despite the shrinkage of the EU share in the relevant world markets, due to the development of the leather industry in other regions of the world, such as Turkey, China, India, Pakistan, Brazil and Ethiopia [1]. Tanning is an important process for transforming rawhides into several leather goods, which are used daily by the consumers. The process of turning hides into leather can be divided into four subsequent treatment phases, i.e., beamhouse operation, tanyard process, retanning and finishing [2]. Nevertheless, for each end-product (e.g., shoes, jackets, bags, couches, chairs etc.), the relevant tanning process is rather specific and the kind and amount of the respectively produced wastes may vary significantly. However, an average amount of wastewater in the range of 30–35 m^3 is usually produced per ton of raw processed material, noting that acids, alkalis, chromium salts, tannins, solvents, sulfides, dyes, auxiliaries and many other chemical compounds, which are used during the processing of leathers, are not completely used/removed with the treated items and, therefore, remain in the produced effluents.

The characteristics of tannery wastewaters can differ significantly among various tannery units, depending on the size/capacity of the specific industry (usually small- and medium-sized enterprises, SMEs), as well as on the applied chemicals, the amount of water used and the type of final product. The presence of several substances can increase substantially the values of fundamental pollution parameters, such as chemical oxygen demand (COD) (average concentration 6,200 mg/L), TDS (average total dissolved solids concentration 87,000 mg/L) etc. [3]. The high COD and SS (suspended solids) loadings in these wastewaters can pose an important economic problem for tanneries, since these parameters have been extensively used by most water/wastewater companies as major indices for the effluent quality, and thus, they are frequently controlled [4].

Therefore, specific consideration should be addressed to re-evaluate the physico-chemical and bacteriological quality of tannery effluents prior to their disposal in the aqueous environment. Various processes have been commonly implemented to treat wastewater from tannery industries, such as biological [3,5–9], oxidation [10–12], chemical processes including coagulation/flocculation [13,14] etc. Additionally, membrane technologies, such as ultrafiltration (UF) and reverse osmosis (RO), can be effectively combined with the conventional tannery wastewater treatment processes to improve their efficiencies. However, the use of membranes for this application was rather limited, at least until few years ago, due to the relatively higher cost of associated capital equipment and consumables. Nevertheless, there has recently been a significant reduction in the cost of membrane systems, which is probably due to the development of more efficient manufacturing processes and the increased competition in the respective market, making the use of membranes for industrial wastewater treatment processes more attractive.

Relevant experiments have been conducted since the late 1950s, which included mostly membranes of natural origin. Following nearly 60 years of rapid advancement, today the membrane-based processes are already implemented in numerous industrial applications, presenting substantial benefits. Cassano et al. [15] described the use of nanofiltration (NF) in order to improve chromium recovery from spent chromium tanning baths and RO to desalinate the water discharged from filter presses after Cr(III) precipitation. The quality of produced RO permeate was satisfactory, for being re-used in washing operations. Suthanthararajan et al. [16] studied a pilot-scale membrane system (with capacity 1 m^3/h), which comprised NF and RO membrane units and several pre-treatment operations. When using the RO membrane, the maximum TDS removal rate was more than 98% and the permeate recovery rate was about 78%. This permeate was shown to have very low TDS concentration and may be reused for the wet finishing process in tanneries. Mendoza-Roca et al. [17] studied the reuse of UF permeate on the quality of final leather, as well as the comparison among different types of membrane cleaning procedures. The results showed that the final quality of the skin was not affected by the use of UF permeate for the unhearing process.

Bhattacharya et al. [18] studied the treatment of high-strength tannery wastewater (COD 5,680 mg/L and BOD 759 mg/L) by using ceramic microfiltration (MF) membranes. This study proposed a two-step treatment unit, which involved MF followed by RO. The treated water was appropriate for reuse in the tanning process, as the values of organic parameters in the effluent were found below the respective control concentrations. Kaplan et al. [19] investigated the treatment of highly polluted tannery wastewater by using three different ceramic MF and UF membrane modules in a cross-flow lab-scale unit. The wastewater samples were received from a tannery outflow in the industrial area of Isparta (Turkey). During these experiments the permeate flow was reduced, although the cake layer on the surface of membrane was appropriately removed by the application of chemical cleaning procedures. Despite the fouling issues, the membranes were able to achieve 95% color removal, while COD removal rate ranged between 58% and 90% for all the applied pressures. Rambabu and Velu [20] investigated the treatment of tannery wastewater by using modified poly-ether-sulfone (PES) membranes. The permeability was significantly increased after the use of modified membranes, although the removal rates of BOD$_5$, COD, TDS, chlorides, sulfates, oils and fats ranged in rather moderate levels.

The vibratory shear-enhanced process (VSEP) is a membrane-separation technology, which was invented in 1987 and patented in 1989. VSEP applies vibration to a membrane in order to increase the separation efficiency and reduce membrane fouling. The application of high shear stresses on the membrane surface results in the removal of the greater part of solids and foulants, i.e., of the substances which are primarily responsible for membrane fouling. It has to be noted also that depending on the applied pressure and the filtration rate, the thickness of cake layer formation can vary. Membranes with different pore diameters/operations (i.e., MF, UF, NF and RO) have been considered for application in the VSEP system. The plate-and-frame configuration, which is the simplest module for packing flat sheet membranes, is typically used in most setups [21]. VSEP filtration has been previously applied in several cases, such as for the treatment of landfill leachates [22], the removal of humic acids in the presence of inorganic particles (clays) from synthetic aqueous dispersions, the treatment of simulated/contaminated surface waters [23], the purification of pulp and mill paper re-circulation water, the treatment of yeast dispersions and bovine albumin solutions, the process of dairy waters and the separation of casein micelles from skimmed milk [24–27].

The objective of the present study was to evaluate and compare the performance of a VSEP lab-scale unit, when using different membrane types (MF, UF and RO), in terms of major contaminants removal during the treatment of simulated or industrial tannery wastewaters.

2. Materials and Methods

2.1. Simulated Tannery Wastewater

The synthetic tannery wastewater (COD 2,000 mg/L) was prepared by dissolving 1.5 g/L tannic acid, 7.0 g/L sodium chlorate, 8.0 g/L sodium sulfate, 2.0 g/L ammonium chloride and 0.1 g/L sodium dodecyl sulfate (SDS) in the tap water of Thessaloniki city (Greece) [28]. These reagents (obtained from Panreac Chemical Company, Barcelona, Spain) were stirred by means of a mechanical agitator for 1 h in a feed tank, made from high-density poly-ethylene (PE). Real tannery wastewaters were received from a nearby medium-sized industrial plant, located at the major industrial area of Sindos (Thessaloniki, Greece).

2.2. Membrane Types

Nine different membranes were evaluated in this study, i.e., three MF membranes (made of Teflon), four UF membranes (made of regenerated cellulose) and two RO membranes (made of polyamide/polyester).The membranes used were flat-disks with an effective membrane area of about 0.05 m^2. For the membrane support a polycarbonate track-etched drainage cloth, made of polypropylene, was also applied. The specific membrane characteristics, as provided by the manufacturing company, are summarized in Table 1. Prior to each experiment, the pure water flux (PWF) was determined for all these membranes in order to evaluate and compare better the performance of each membrane; the experimentally determined values were found to be consistent with the reported values from the manufacturing company.

2.3. Analytical Methods

The permeate flow rate was recorded at certain time intervals during each experiment with a calibrated volumetric cylinder and a timer. Moreover, a thermocouple was used into the feed tank in order to monitor the temperature throughout the experiments. The concentration of humic substances (used as common representative of natural organic matter, NOM) in the samples was estimated with a Shimadzu spectrophotometer (Shimadzu Instruments, MD, USA) after measuring the ultraviolet (UV) absorbance at 254 nm. Moreover, the removal of tannins, NOM and aromatic compounds was estimated from the reduction of UV absorbance measurements at 700, 254 and 220 and 275 nm, respectively. All samples were measured at natural pH value, i.e., without adjustment, after the addition of the respective sample into a quartz cell and the comparison with another cell, which contained deionized water and was used as a reference.

The color was measured at the wavelength of 455 nm, according to the Standard Methods for the Examination of Water and Wastewater (APHA) (1992), which were also applied for the determination of COD, N_{total}, NO_3^- and NH_4^+ [29]. Color was measured with a HACH spectrophotometer at 455 nm, according to the respective APHA guideline (Method No. 2120C), and was reported as platinum–cobalt (PtCo) units, i.e., the color produced by 1 mg Platinum/L in the form of chloro-platinate complex ion. The pH values of feed and permeate were measured with a pH meter (Jenway, model 3540, Essex, UK). The suspended solids concentration was evaluated by using a Hachturbidimeter (Hach), while the total organic compounds were measured by a total organic carbon (TOC) analyzer (Shimadzu, MD, USA).

Table 1. Major physico-chemical properties of the nine membranes examined in the present study, as indicated by the manufacturer of the vibratory shear-enhanced process (VSEP) treatment system (New Logic International, Concord, USA).

Number Code	T-0.1	T-0.45	T-1.0	C-200	C100	C-30	C-10	TFC-99	TFC-96
Process	MF	MF	MF	UF	UF	UF	UF	RO	RO
Material	PTFE	PTFE	PTFE	Regenarated Cellulose	Regenarated Cellulose	Regenarated Cellulose	Regenarated Cellulose	Polyamide Polyester	Polyamide Polyester
Cut off-diameter (μm for MF and kDa for UF)	0.1	0.45	1.0	200	100	30	10	Rej. NaCl 99%	Rej. NaCl 96%
Maximum operating pressure (bar)	7	7	7	20	20	20	20	40	40
pH range (20 °C)	2–11	2–11	2–11	2–11	2–11	2–11	2–11	2–11	2–11
Pure water flux (PWF) at max operating pressure (L m^{-2} h^{-1})	600	600	750	1050	1000	100	80	100	100

The removal efficiency of a specific component (pollutant) by any membrane is defined as:

$$R(\%) = \frac{C_o - C_p}{C_o} \times 100\% \quad (1)$$

where R is the removal efficiency of the membrane for a given pollutant at a defined hydrostatic pressure and feed concentration, whereas C_p and C_o are the concentrations of the rejected components in the permeate and in the feed, respectively.

2.4. Vibratory Shear-Enhanced Process (VSEP) Module

The VSEP membrane filtration module (Figure 1) was a relatively small pilot-scale module, manufactured by New Logic International (USA), and its characteristics and operation has been previously described in detail [23]. The module consisted of an annular membrane (2) with an area of 503 cm^2, in a circular housing (1), placed at the top of a vertical shaft (3), acting as a torsion spring. This shaft amplifies the vibrations, which are generated at the bottom plate by an eccentric drive motor (4). The membrane oscillated azimuthally in its own plane with the examined amplitudes being 0, 6.35, 12.7, 19.05, 25.4 and 31.75 mm. The shear stress, which is created on the surface of membrane, is produced by the inertia of the fluid, as in the case of Stokes layer near an oscillating plate [24–26].

The VSEP module was fed by a volumetric diaphragm pump (5) from a stirred tank, containing the test feed/tannery wastewater in this case. Permeate was received from a valve at the top of the housing (6) and collected into a beaker at atmospheric pressure. The concentrated slurry was returned through the "process out" line, as shown in Figure 1. The return flow passed through the flow limiter (7) and the control valve (8), which allowed the fine adjustment of outlet pressure. Inlet and outlet pressures were measured by Validyne analog gauges in order to determine the trans-membrane pressure (TMP).

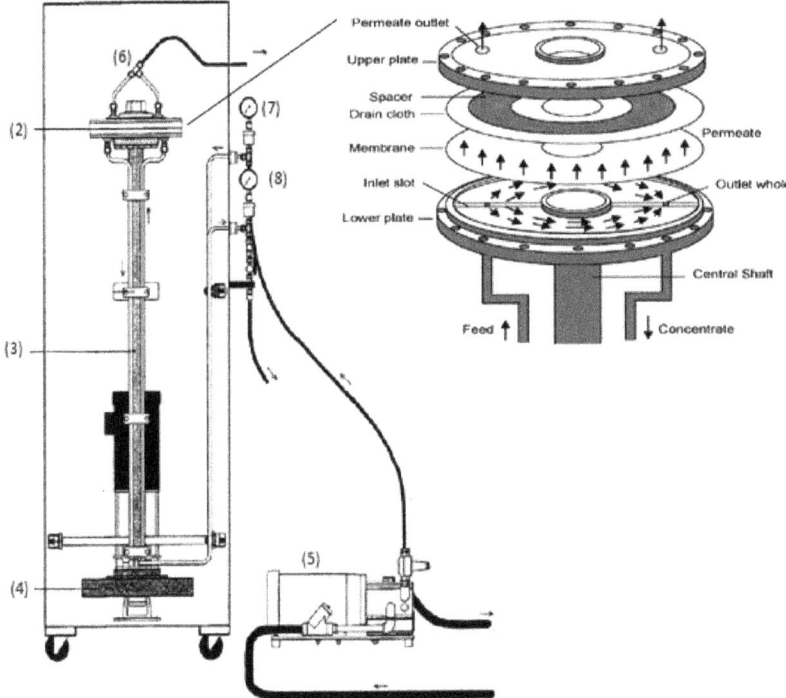

Figure 1. VSEP module: (1) circular housing, (2) annular membrane, (3) vertical shaft, (4) eccentric drive motor, (5) volumetric diaphragm pump, (6) top of the housing, (7) flow limiter, (8) control valve.

3. Results and Discussion

3.1. Microfiltration (MF) Experiments

During the MF experiments, three membranes (made of teflon) were tested with average pore diameters 0.1, 0.45 and 1.0 μm. The TMP was kept constant at around 5 bars (as recommended by the manufacturer), while the vibration amplitude varied from 6.35 to 31.75 mm.

Filtration is characterized as "dead-end", when the flow is applied perpendicular to the membrane surface. Particles that are smaller than the effective membrane pore size can pass through the membrane as permeate, while particles that are larger build up and result in the formation of a cake layer on the membrane surface. Filtration is characterized as "cross-flow", when the flow is applied tangentially across the membrane surface. In this case, as the feed flows across the membrane surface, the permeate passes through the membrane, while the concentrate or retentate accumulates at the opposite side of the membrane. This tangential flow creates a shear stress on the surface of the membrane, which in turn can reduce fouling. Because the cross-flow operation is capable of removing most of the formed cake layer from the surface of membrane, the permeate flux does not decrease as fast as in the case of dead-end filtration. Cross flow filtration also offers the benefit of an increased membrane lifespan by preventing irreversible fouling.

In order to examine the influence of VSEP system on the treatment of tannery wastewater, 50 L were fed into the membrane module. Permeate was discharged, while the retentate during the first series of experiments was thrown out (i.e., using a "dead-end" operational mode). By contrast, during the second series of experiments, the retentate was re-circulated to the feed tank (i.e., using a "cross-flow" operational mode) and, therefore, led to the gradual increase of pollutants' content with the increase of treatment (operational) time, i.e., the semi-batch operation mode was employed in the latter case. TMP was maintained constant and the permeate flow rate was recorded at regular intervals. The final tannery wastewater volume after each filtration experiments (by applying the semi-batch treatment) was equal to 0.5 L, resulting in 95% of volume recovery for the case of MF. The rejection (removal) of tannins, NOM and aromatic compounds was estimated from the reduction of UV absorbance measurements at 700, 254 and 220 and 275 nm, respectively, while the rejection of particles was estimated from the respective turbidity measurements.

In all cases it was observed that the permeate flux was stabilized after about 20 min of filtration. The results of MF by using the 0.45 μm membrane after 30 min of VSEP system operation are presented in Table 2.

Table 2. Removal (%) of major pollution parameters (total organic carbon (TOC), chemical oxygen demand (COD), N_{total} and tannins), when the microfiltration (MF) process was applied for the treatment of simulated tannery wastewaters.

Pressure 5 Bar	TOC	COD	N_{total}	Tannins
Vibration amplitude				
0	15	25	23	25
6.35	26	26	27	32
12.7	35	27	28	36
19.05	45	38	32	37
25.4	48	45	38	38
31.75	52	55	42	40

Although MF is usually applied for the rejection of suspended solids, it was found that the TOC, COD, N_{total} and tannins removal rates were also quite significant in dead-end (DEP) as well as in cross-flow (CFP) operating modes (see Table 2 and Figure 2). Moreover, it was found that the higher amplitude of vibration affected positively not only the quality of the filtered wastewater, but also the feeding rate of the system (data not presented), in all the examined cases of MF membranes. Specifically, at the maximum vibration amplitude, the lower concentrations of pollutants in the produced filtrate were determined (with R_{TOC} = 52%, R_{COD} = 55%, R_{Ntotal} = 42%, $R_{tannins}$ = 40%). The same conclusion was reached by other investigators as well [30], who treated different wastewaters with the VSEP system, comparing two vibrational amplitudes, of which the maximum one produced the best treatment results.

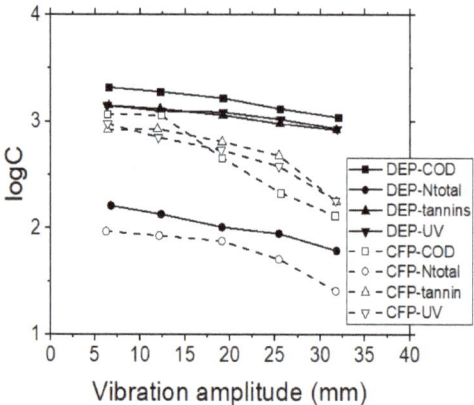

Figure 2. Effect of vibration amplitude on COD, N_{total}, ultraviolet (UV) 254 nm and tannins concentrations, when performing MF membrane experiments, during dead-end (DEP) or cross-flow (CFP) operating modes.

In addition, more experiments were performed with wastewaters having different (initial) concentrations of pollutants, i.e., lower (7 mg/L) or higher (50 mg/L) total nitrogen content. In the experiments with the lower initial nitrogen concentration (7 mg/L), the removal by the MF membranes was 75%, while with the higher initial nitrogen concentration (50 mg/L) the removal was only 25%. Consequently, the MF is particularly advantageous for the removal of particulate matter (turbidity), while it is not so effective for the removal of soluble pollutants (e.g., N_{total}). The removal efficiency of particulate matter, when using the 0.45 μm membrane, is 20% higher than that of the 1.0 μm membrane and 20% lower than that of the 0.1 μm membrane, respectively. Microfiltration, by using the membrane with pore diameter 0.1 μm, gave the best removal results of particulate pollutants, as well as of the other studied parameters, such as turbidity, UV_{254nm} absorption, color and tannins.

However, it is expected that as the pore diameter of the membrane decreases, the permeate flow rate is also reduced. In the present study this behavior was consequently observed, i.e., the membrane having the largest pore diameter in comparison with the membrane with the smallest pore diameter. The 0.45 μm membrane exhibited 20% lower flow rate than the 1.0 μm membrane, and the 0.1 μm presented 20% and 40% lower flow rates than the 0.45 μm and 1.0 μm membranes, respectively (Figure 3). On the contrary, it was observed that the permeate flow decreased as the operational time increased, due to the creation of a fouling layer on the membrane surface. In the particular vibrating membrane system, it was observed that the permeate flow rate was reduced for a period of about 1 h and then it was stabilized until the end of the experiment (Figure 3), i.e., the permeate flow rate was found to change over (operation) time. A power regression model, which is of the general form $y = ax^b$, was used to describe this behavior, indicated as the solid lines in Figure 3. The b factor for the three examined membranes (1.0, 0.45 and 0.1 μm as "nominal" pore diameters) was defined as -0.101, -0.120 and -0.090, respectively.

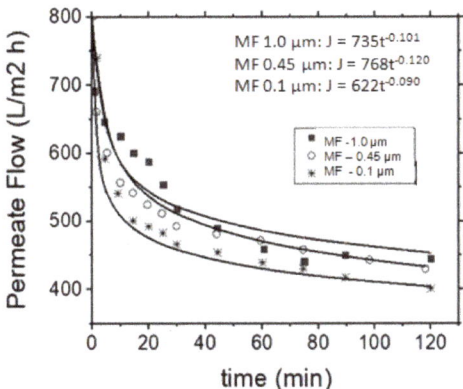

Figure 3. Permeate flow rate variation (or flux) vs. operation time, regarding the application of MF process for the treatment of simulated tannery wastewaters.

3.2. Ultrafiltration (UF) Experiments

In this case the effectiveness of four different membranes, made from regenerated cellulose, was evaluated. The regenerated cellulose is more hydrophilic and, therefore, suitable for the separation of respective compounds, such as carbohydrates or tannins. Dissolved organics, e.g., humic acids, proteins, carbohydrates and tannins, are the most serious foulants and they are more difficult substances to remove. Hydrophilic membranes have been found less prone to fouling by organic colloids [31,32]. Preliminary experiments were initially carried out for each membrane type in order to determine the critical TMP values within the respective pressure range as recommended by the manufacturer. During the UF experiments, the TMP was raised every 10 min, until the permeate flux reached an equilibrium stage, where it became independent of TMP. This critical TMP value was selected as the operational pressure and was determined at 14 bars (Figure 4).

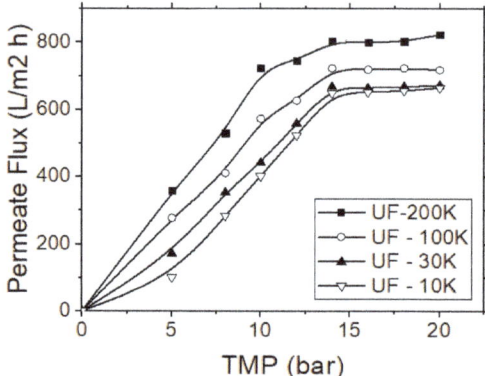

Figure 4. Permeate flux variation vs. the applied trans-membrane pressure (TMP) for the application of the ultrafiltration (UF) process, regarding the treatment of simulated tannery wastewater.

The treatment of (simulated) tannery wastewater by using UF membranes was found to produce better results regarding the removals of organics, nitrogen, turbidity and tannins content when compared with the previous MF experiments. Figure 5 shows the (total) nitrogen removal rate for each membrane type, starting from the membrane with the largest molecular weight cut-off (MWCO) to the membrane with the smallest one (i.e., 200, 100, 30 and 10 MWCO). Nitrogen removal during the UF experiments was not found particularly high, although it surpassed the respective values of MF process, reaching 75% by applying the optimal pressure (14 bars) and the highest vibration amplitude. Zouboulis and Petala [23] observed a slight increase of humic substances removal rate when increasing the vibration amplitude (i.e., the shear rate on the membrane surface). The increase of vibration amplitude can induce higher shear stresses on the surface of membrane, which corresponds to an increase of particle–particle collisions; thus, to an increase of shear diffusion that forces the particles away from the membrane surface and back to the bulk solution. Furthermore, the reduction of concentration polarization in this case is expected also to lower the concentration of contaminants at the membrane surface and their diffusive transfer through the membrane.

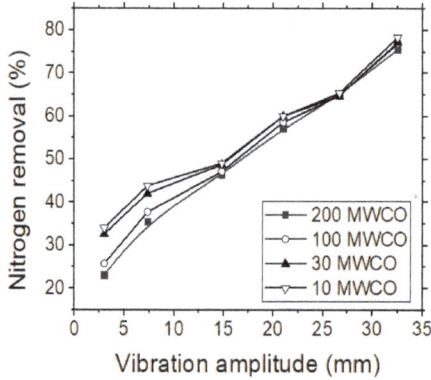

Figure 5. VSEP treatment of simulated tannery wastewaters; the effect of vibration amplitude on the N_{total} removal (%) rate by applying four different UF membranes.

Moreover, in all these examined cases the permeate flux decreased abruptly during the initial 15 min of each experiment (Figure 6), and stabilized after 40 min of operational time, while the retentate stream was continuously re-introduced into the feed tank (following the semi-batch operational mode). The application of 30 and 10 kDa MWCO membranes during the UF experiments, showed that the 30 kDa membrane is more efficient, although the removal capacities of examined pollution parameters for both membranes were rather similar, but the 30 kDa membrane enabled higher permeate fluxes. Figure 6 depicts the variation of permeate flux with the operational time for the used four different UF membranes. The permeate flow rate, as in the case of MF, is linked to the respective time. The b factor of the aforementioned power regression model for the 200 and 100 kDa membranes is almost equal (−0.077 and −0.076) and the same was observed for the cases of 30 and 10 kDa membranes (−0.096 and −0.092).

Considering the quality of permeate, tannins were rejected by over 85% during the UF process, when using the 200 or 30 kDa membranes, while their removal was even higher than 92% by using the 10 kDa membrane (Figure 7). Both permeate flux and permeate quality was stabilized for all the examined cases after the first 40 min of operational time. In addition, the experimental data indicate that tannins' rejection rate was almost independent from the initial feed concentration; thus, it was considered that the retention of tannins was depended mostly on the specific MWCO membrane used, which in turn can affect the steric hindrance and the adsorption capacity, as well as the porosity of the formed cake layer near the membrane surface. It can be assumed that in these cases, the fraction of tannins which passed through the membrane was of lower molecular weight.

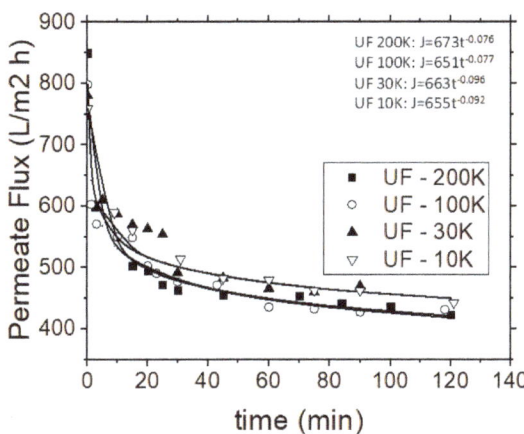

Figure 6. Permeate flow rate variation vs. operational time for the UF process.

3.3. Reverse Osmosis (RO) Experiments

The efficiency of RO process on tannins and organic content removal was also examined, by using a polyamide/polyester membrane. The TMP was adjusted to 20 bars after preliminary experiments, which were performed in order to determine the critical operational TMP within the respective pressure range, as recommended by the manufacturer. The flux decline was recorded every 3 min during each experiment. The permeate flux was initially influenced by the feed quality and the filtration operational time, when the system operated in the semi-batch mode (i.e., by applying the retentate re-circulation), but it reached an equilibrium stage after 20 min of operation. The results obtained by the RO membranes, with respect to vibrational amplitude and permeate flux, corresponded with the previously examined MF and UF membranes (data not shown). The removal capacities of contaminants by using the RO membranes for dead-end and cross-flow operation modes are depicted in Figure 8. In all cases, the cross-flow process shows better results, both for the quality of permeate, as well as for the hydrodynamic behavior of the examined system.

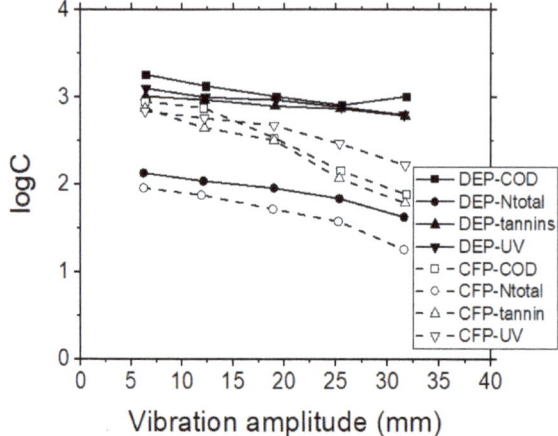

Figure 7. VSEP treatment of simulated tannery wastewaters; the effect of vibration amplitude (mm) on COD, N_{total}, UV 254 nm and tannins residual concentrations, when performing UF membrane experiments, by applying dead-end (DEP), or cross-flow (CFP) operational modes.

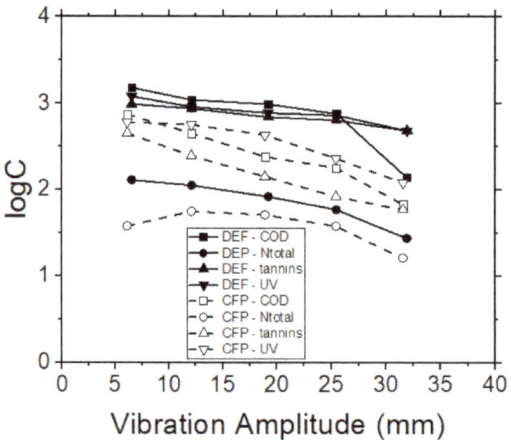

Figure 8. VSEP treatment of simulated tannery wastewaters; effect of vibration amplitude on COD, N_{total}, UV 254 nm (NOM) and tannins concentrations, when performing RO membrane experiments, during dead-end (DEP) or cross-flow (CFP) operational modes.

3.4. Comparing the Membrane Processes by Treating Real Industrial Tannery Wastewater

Relevant membrane filtration experiments were subsequently conducted, by using real industrial tannery wastewater in dead-end and cross-flow operation modes and by applying 3 membrane processes (MF, UF and RO). During the dead-end flow mode the raw water feed passes directly through the membrane, in contrast to the cross-flow filtration mode, which employs a high velocity of the raw water feed, flowing in parallel over (and across) the membrane surface. The TMP was kept constant and the permeate flow rate was recorded at regular time intervals. Vibration amplitude was set to 0.025 mm, according to preliminary experiments. The permeate fluxes were stabilized after a certain period of time and the system reached equilibrium conditions after approximately 50 min of filtration time in dead-end mode and after 60 min in cross-flow modes (Figure 9). This was attributed to the capability of the VSEP unit to develop high shear stresses on the membrane surface due to vibration.

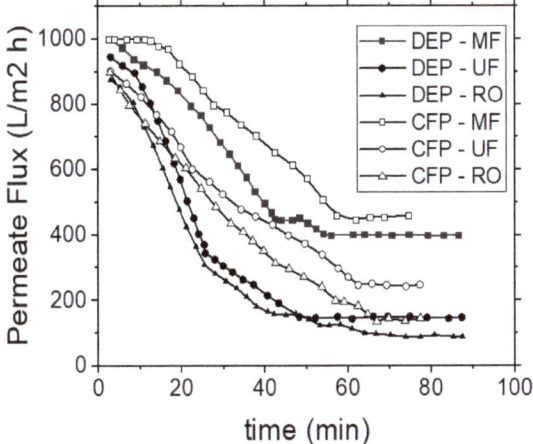

Figure 9. Treatment of real industrial tannery wastewaters; permeate flux variation vs. operational time during dead-end (DEP) or cross-flow (CFP) operation modes.

Regarding the results obtained (quantitative removal of major pollution parameters) by the VSEP treatment system, they can be considered as satisfactory (Table 3), considering that they were obtained by applying a single (one-step) direct treatment process. In combination with appropriate pre- or post-treatment techniques (biological or physico-chemical, according to the relevant literature) further improvement of these results can be expected.

Table 3. Characteristics of real industrial tannery wastewater before and after the application of the VSEP treatment process.

Parameter	Industrial Tannery Wastewater Feed	Industrial Tannery Wastewater after VSEP Treatment
COD (mg/L)	7500	950
N_{total} (mg/L)	1055	550
$N-NH_4^+$ (mg/L)	4.6	2.6
$N-NO_3^-$ (mg/L)	20	5.6
Turbidity (NTU)	>2000	250

3.5. Theoretical Considerations and Calculation of Mass Transfer Coefficient

Initially, the effect of vibration amplitude on the permeate flow rate was examined without the recirculation of concentrate. Thus, according to Equations (2) and (3) [24–26], the average and maximum shear rate was determined (Table 4):

$$\gamma_{w,max} = \frac{R_2 \cdot \Omega \cdot Re^{\frac{1}{2}}}{h} = (2\pi F)^{\frac{1}{2}} R_2 \Omega v^{\frac{-1}{2}} = 2^{\frac{1}{2}} d (\pi F)^{\frac{3}{2}} v^{\frac{-1}{2}} \tag{2}$$

$$\gamma = \frac{2^{\frac{3}{2}} (R_2^3 - R_1^3)}{3\pi R_2 (R_2^2 - R_1^2)} \gamma_{w,max} \tag{3}$$

where d is the peak to peak vibration amplitude at the periphery of the membrane (m), F is the vibration frequency (Hz) and ν is the kinematic viscosity of the fluid (m^2 s^{-1}).

Table 4. Average and maximum shear rate of the studied VSEP system.

Frequency (Hz)	Vibration Amplitude (m)	γ_{max} (s^{-1})	γ_w (s^{-1})
53.52	0.0064	19,564	6397
54.30	0.013	39,986	13,076
54.60	0.019	60,636	19,828
54.76	0.025	78,122	25,546

The average and maximum shear rates are proportional, depending on the amplitude of the membranes' vibration (Figure 10), whereas the permeate flow rate increases exponentially and in relation to the vibration amplitude and to the average shear rate. The relationship between the permeate flow rate with the vibrational amplitude and the shear rate was also determined, according to Equations (4) and (5), and the results obtained were found to be consistent with the relevant published research studies [24–26,33].

$$J = 407 d^{0.77} \quad (4)$$

$$J = 0.37 \gamma^{0.44} \quad (5)$$

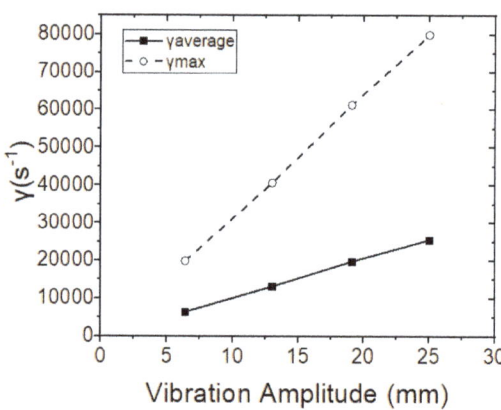

Figure 10. Average and maximum shear rates in relation to vibration amplitude.

The mass transfer coefficient was subsequently determined in the synthetic and industrial tannery wastewaters by using the 10K UF membrane. The mass balance concept in the membrane filtration processes can be described with the following equation:

$$\frac{C_m - C_p}{C_b - C_p} = \exp\left(\frac{J_v}{k}\right) \quad (6)$$

where C_p is the concentration of solute in the permeate, C_m is the concentration of solute on the membrane surface, C_b is the concentration of solute in the bulk, J_v is the volumetric flux of pure water and k is the mass transfer coefficient.

The rejection characteristics of membranes are further discussed, by using the observed rejection rate ($R_{obs.}$), as well as the real rejection rate (R_{real}), according to Equations (7) and (8) [34]:

$$R_{obs.} = \frac{C_b - C_p}{C_b} \quad (7)$$

$$R_{real} = \frac{C_m - C_p}{C_m} \quad (8)$$

By using these equations, Equation (6) can be rewritten as:

$$\ln\left(\frac{1-R_{obs.}}{R_{obs.}}\right) = \ln\left(\frac{1-R_{real}}{R_{real}}\right) + \frac{1}{\gamma}\left(\frac{J_v}{u^a}\right) \quad (9)$$

In this Equation (9) a reasonable value for α can be set and then a linear relationship between $\ln\left(\frac{1-R_{obs.}}{R_{obs.}}\right)$ and $\left(\frac{J_v}{u^a}\right)$ can be obtained. R_{real} can be obtained by extrapolating the linear plots of $\ln\left(\frac{1-R_{obs.}}{R_{obs.}}\right)$ vs. $\left(\frac{J_v}{u^a}\right)$ and then, k is obtained from the following equation:

$$k = \gamma u^a \quad (10)$$

From the slope of the straight line connecting the average shear rate and the $\ln\left(\frac{1-R_{obs.}}{R_{obs.}}\right)$ term, the mass transfer coefficient can be subsequently determined. For the experiments using synthetic tannery wastewater the mass transfer coefficient was found to be 1.7×10^{-3}.

3.6. Finding the Main Membrane Fouling Mechanism

The mathematical modeling of flux decline during membrane filtration can provide a better understanding of membrane fouling, as well as contribute to appropriate predictive tools for the successful scale-up or scale-down of filtration systems. The main empirical models, which are used in order to explain the permeate flux behavior and to determine the involved (main) fouling mechanisms are the Hermia models [35]. Hermia developed four empirical models which include four major types of fouling, i.e., (i) the complete pore blocking, (ii) the intermediate blocking, (iii) the standard blocking, and (iv) the cake layer formation. The parameters of these models represent a physical meaning and correspond to the respective fouling mechanism(s) [36]. Hermia's models were originally developed for the dead-end filtration operational mode and were based on the constant pressure filtration laws. However, despite the different sets of applied mass and momentum equations for dead-end and cross-flow filtration operation models, several researchers have applied Hermia's models to describe also the cross-flow filtration operational mode.

Hermia's model is expressed by the following general differential Equation (11):

$$\left(\frac{d^2 t}{dV^2}\right) = K\left(\frac{dt}{dV}\right)^n \quad (11)$$

Noting that V is the accumulated permeate volume (m^3), t is the filtration time (s) and K and n are the phenomenological coefficient and the general index, respectively, both depending on the type of fouling (K is a unit dependent on the parameter n in Equation (11)). In the following sections the aforementioned models will be shortly presented.

3.6.1. Complete Pore Blocking Model ($n = 2$)

When the particles to be separated are larger than the membrane's pore size, then there is a pore blockage, due to pore obstruction and sealing. Hermia concluded that in this case the parameter n is equal to 2. For $n = 2$, Equation (11) is expressed in terms of permeate flux vs. time, according to the following equation [37]:

$$\ln(J_p) = \ln(J_0) - K_c t \quad (12)$$

Noting that J_p is the permeate flux (L/m^{2}h), J_0 is the initial permeate flux (L/m^{2}h) and K_c (m^{-1}) is the equation constant.

The parameter K_c can be described as a function of blocked membrane surface, per unit of total permeate volume K_A, and as a function of the initial permeate flux J_0, as shown in Equation (13) [38]. As a result, the active membrane area is reduced; due to the pores being completely blocked [39].

$$K_c = K_A J_0 \tag{13}$$

3.6.2. Standard Blocking Model ($n = 3/2$)

When the solute's molecular size is smaller than the membrane pore size, then the pore blocking possibly occurs inside the pores [40]. This model considers that the separated particles can be either adsorbed or deposited on the walls of the membrane's pores. Therefore, the available (free) volume of membrane pores decreases proportionally to the permeate volume, which passes through the membrane. As a result, the cross sectional area of the membrane pore decreases with time, and consequently the membrane resistance increases [39]. It is considered that the pores' lengths and diameters are relatively constant along the entire membrane surface. Considering these hypotheses, Hermia [35] concluded that the parameter n is equal to 3/2 in this case. Considering the respective blocking (fouling) mechanism, the permeate flux can be expressed as a function of time, according to Equation (14):

$$\left(\frac{1}{J_p^{\frac{1}{2}}}\right) = \left(\frac{1}{J_0^{\frac{1}{2}}}\right) + K_s t \tag{14}$$

The parameter K_s can be calculated, according to Equation (15):

$$K_s = 2\frac{K_B}{A_0} A x J_0^{\frac{1}{2}} \tag{15}$$

Noting that K_B is a parameter that represents the decrease of cross-sectional area of membrane pores per unit of total permeate volume (s^{-1}), J_0 is the initial permeate flux (L/m²h), A is the membrane surface (m²) and A_0 is the membrane porous surface (m²).

3.6.3. Intermediate Blocking Model ($n = 1$)

When the size of particles is similar to the membrane's pore size, the intermediate blocking mechanism may take place. As in the case of complete pore blocking model, this model considers that solid particles (or even macromolecules) that at any time reach an open pore, might block it. Nevertheless, a dynamic situation of the blocking/unblocking state may also occur. Also, the particles may bridge a pore by blocking the opening, but not completely seal it [39]. Considering these hypotheses, Hermia [35] concluded that the parameter n in this case is equal to 1. Other researchers [40] expressed the permeate flux as a function of time, resulting in Equation (16):

$$\frac{1}{J_p} = \frac{1}{J_0} k_i t \tag{16}$$

Noting that J_p is the permeate flux (L/m²h) and J_0 is the initial permeate flux (L/m²h), the parameter K_i (m^{-1}) can be expressed as a function of blocked membrane surface per unit of total permeate volume, i.e., as K_A (Equation (17)). The area of membrane surface that is not blocked diminishes with time [41]. As a result, the probability of a molecule blocking/fouling a membrane pore is continuously decreasing with time.

$$K_i = K_A \tag{17}$$

3.6.4. Cake Layer Formation Model ($n = 0$)

As in the case of pore blocking model, in this case the solute molecules are larger than the membrane pore size, and they cannot penetrate through them [41]. In this model, a cake layer is formed on the surface. Nevertheless, when the concentration of solute molecules is considerable, they can be deposited on the surface or on the previously deposited layers, resulting in the formation of multiple layers. For the cake layer formation model, the permeate flux is given as a function of time by the (linearized) Equation (18):

$$\left(\frac{1}{J_p^2}\right) = \left(\frac{1}{J_0^2}\right) + K_{gl} t \tag{18}$$

The parameter K_{gl} can be defined, according to Equation (19):

$$K_{gl} = 2 \frac{K_D x R_g}{J_0 x R_m} \tag{19}$$

Noting that K_D represents the cake layer area per unit of total permeate volume (1/m^3), R_g is the cake layer resistance (m^{-1}) and R_m is the hydraulic membrane resistance (m^{-1}).

3.6.5. Application of Hermia's Model for the Indication of Major Fouling Mechanism, When Treating Simulated or Real Industrial Tannery Wastewater by the VSEP System

Figure 11 illustrates the results obtained after processing the experimental data for the synthetic, as well as for the real industrial tannery wastewater. The slope of the straight line in the case of synthetic and of industrial tannery wastewater was 0.88 and 0.68, respectively. Therefore, the intermediate blocking model describes the results of the present research better in comparison with the other models.

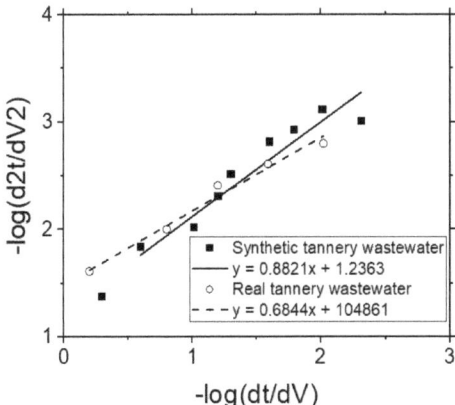

Figure 11. Presentation of the experimental data following the Hermia's model in order to find the (major) mechanism of membrane fouling during the application of VSEP for the treatment of simulated or real industrial wastewater.

4. Conclusions

The aim of the present study was the treatment of simulated and industrial tannery wastewater by using a vibratory shear-enhanced process (VSEP) system. A variety of parameters affecting the rejection efficiency of pollutants were studied, such as the specific membrane separation process (MF, UF, RO), the different membrane type for each case, and the applied vibration amplitude. The main conclusions of the study are the following:

- As the vibration amplitude increased, the respective fouling phenomena were restricted.
- The system's hydrodynamic behavior was satisfying, because the permeate flux remained almost constant, even during the cross-flow filtration mode, and even when the feed stream contained progressively higher concentrations of contaminants, due to the recirculation of retentate in the feed tank.

- A higher rejection rate was observed during the UF process, by using the 10 K membrane, as well as during the RO process.
- The rejection of components/pollutants increased with the increase of vibration amplitude, due to the enhancement of shear diffusion.
- The organic matter removal in terms of COD values exceeded 75% for all the examined cases. Furthermore, UF membranes had similar COD removal rates (about 80%–87%), while the MF membrane retained 65% COD and the RO was even more efficient (reaching up to 96% COD removal).
- Although the operation of VSEP and the application of higher TMPs during the MF and UF processes enhanced the membrane-fouling mitigation, the operation of the treatment system was still satisfactory.
- The theoretical calculations showed that the most likely fouling mechanism is the intermediate blocking, according to Hermia's classification system.

Author Contributions: Conceptualization, A.I.Z. and E.N.P.; methodology, A.N. and A.I.Z.; validation, A.I.Z., E.N.P. and A.N.; formal analysis, E.N.P. and A.N.; investigation, A.I.Z., E.N.P. and A.N.; data curation, A.N.; writing—original draft preparation, E.N.P. and A.N.; writing—review and editing, A.I.Z. and E.N.P.; visualization, E.N.P.; supervision, A.I.Z.

Funding: This research received no external funding.

Acknowledgments: Thanks are due to the Central Tannery Wastewaters Treatment facility in the industrial area at Sindos Thessaloniki (Greece) and especially to N. Apostolidis, P. Karamanolis (Chemical Engineers), D. Liantas and P. Sidiropoulos (technicians) for their valuable help.

Conflicts of Interest: The authors declare no conflict of interest.

Abbreviations

A	Membrane surface (m^2)
A_0	Membrane porous surface (m^2)
C_b	Concentration of the solute in the bulk
C_m	Concentration of the solute at the membrane surface
C_o	Concentration of rejected components in the feed
C_p	Concentration of rejected components in the permeate
d	Peak to peak vibration amplitude at the periphery of membrane (m)
F	Vibration frequency (Hz)
J	Permeate flow rate (L/m^{2}h)
J_0	Initial permeate flux (L/m^{2}h)
J_v	Pure water flux (L/m^{2}h)
k	Mass transfer coefficient
K	Phenomenological coefficient
K_A	Parameter that represents the blocked membrane surface per unit of the total permeate volume (m^{-1})
K_B	Parameter that represents the decrease in the cross-sectional area of the membrane pores per unit of the total permeate volume (s^{-1})
K_c	Constant that corresponds to the complete pore blocking model (m^{-1})
KD	Parameter that represents the cake layer area per unit of the total permeate volume (m^{-3})
K_{gl}	Constant that corresponds to the cake layer formation model (s/m^6)
K_i	Constant that corresponds to the intermediate blocking model (m^{-1})
K_s	Constant that corresponds to the standard blocking model (s^{-3})
n	General index depending on type of fouling
t	Filtration time (min)
R	Percentage removal efficiency of the membrane for a given pollutant at a defined hydrostatic pressure and feed solution concentration (%)
R_{obs}	Observed rejection (%)
R_{real}	Real rejection (%)

R_2	Outer radius of annular membrane (cm)
R_1	Inner radius of annular membrane (cm)
R_g	Cake layer resistance (m^{-1})
R_m	Hydraulic membrane resistance (m^{-1})
u	Water flow velocity inside the fiber (m s^{-1})
V	Accumulated permeate volume (L)
$\gamma_{w,max}$	Maximum shear rate (s^{-1})
γ	Average shear rate (s^{-1})
ν	Kinematic viscosity of the fluid (m^2 s^{-1})

References

1. Orhon, D.; AtesGenceli, E.; Cokgor, E.U. Characterization and modeling of activated sludge for tannery wastewater. *Water Environ. Res.* **1999**, *71*, 50–63. [CrossRef]
2. Lofrano, G.; Meriç, S.; Zengin, G.E.; Orhon, D. Chemical and biological treatment technologies for leather tannery chemicals and wastewaters: A review. *Sci. Total Environ.* **2013**, *461–462*, 265–281. [CrossRef] [PubMed]
3. Durai, G.; Rajasimman, B. Biological treatment of tannery wastewater—A review. *J. Environ. Sci. Technol.* **2011**, *4*, 1–17. [CrossRef]
4. Song, Z.; Williams, C.J.; Edyvean, R.G.J. Characteristics of the tannery wastewater. In Proceedings of the Icheme Research Event, Newcastle, UK, 7–8 April 1998.
5. Sabumon, P.C. Perspectives on Biological Treatment of Tannery Effluent. *Adv. Recycl. Waste Manag.* **2016**. [CrossRef]
6. Ahn, D.H.; Chung, Y.C.; Yoo, Y.J.; Pak, D.W.; Chang, W.S. Improved treatment of tannery wastewater using *zoogloea ranigera* and its extracellular polymer in an activated sludge process. *Biotechnol. Lett.* **1996**, *18*, 917–922. [CrossRef]
7. Wiemann, M.; Schenk, H.; Hegemann, W. Anaerobic treatment of tannery wastewater with simultaneous sulphide elimination. *Water Res.* **1998**, *32*, 774–780. [CrossRef]
8. Farabegoli, G.; Carucci, A.; Majone, M.; Rolle, E. Biological treatment of tannery wastewater in the presence of chromium. *J. Environ. Manag.* **2004**, *71*, 345–349. [CrossRef]
9. Haydar, S.; Aziz, J.A.; Ahmad, M.S. Biological Treatment of Tannery Wastewater Using Activated Sludge Process. *Pak. J. Eng. Appl. Sci.* **2009**, *1*, 61–66.
10. Sivagami, K.; Sakthivel, K.P.; Nambi, I.M. Advanced oxidation processes for the treatment of tannery wastewater. *J. Environ. Chem. Eng.* **2017**. [CrossRef]
11. Schrank, S.G.; Jos, H.J.M.; Moreira, R.F.P.; Schroder, H.F. Fentons oxidation of various—based tanning materials. *Desalination* **2003**, *50*, 411–423.
12. Dogruel, S.; Ates, G.E.; Germirli, B.F.; Orhon, D. Ozonation of nonbiodegradable organics in tannery wastewater. *J. Environ. Sci. Health* **2004**, *39*, 1705–1715. [CrossRef]
13. Shegani, G. Treatment of Tannery Effluents by the Process of Coagulation. *Int. J. Environ. Chem. Ecol. Geol. Geophys. Eng.* **2014**, *8*, 240–244.
14. Chowdhury, M.; Mostafa, M.G.; Biswas, T.K.; Saha, A.K. Treatment of leather industrial effluents by filtration and coagulation processes. *Water Resour. Ind.* **2013**, *3*, 11–22. [CrossRef]
15. Cassano, A.; Criscuoli, A.; Drioli, E.; Molinari, R. Clean operations in the tanning industry: Aqueous degreasing coupled to ultrafiltration. *Clean Prod. Process.* **1999**, *1*, 257–263. [CrossRef]
16. Suthanthararajan, R.; Chitra, K.; Ravindranath, E.; Umamaheswari, B.; Rajamani, S.; Ramesh, T. Anaerobic Treatment of Tannery Wastewater with Sulfide Removal and Recovery of Sulfur from Wastewater and Biogas. *J. Am. Leather Chem. As.* **2004**, *99*, 67–72.
17. Mendoza-Roca, J.A.; Galiana-Aleixandre, M.V.; Lora-Garcia, J.; Bes-Pia, A. Purification of tannery effluents by ultrafiltration in view of permeate reuse. *Sep. Purif. Technol.* **2010**, *70*, 296–301. [CrossRef]
18. Bhattacharya, P.; Roy, A.; Sarkar, S.; Ghosh, S.; Majumdar, S.; Chakraborty, S.; Mandal, S.; Mukhopadhyay, A.; Bandyopadhyay, A. Combination technology of ceramic microfiltration and reverse osmosis for tannery wastewater recovery. *Water Resour. Ind.* **2013**, *3*, 48–62. [CrossRef]

19. Kaplan-Bekaroglu, S.S.; Gode, S. Investigation of ceramic membranes performance for tannery wastewater treatment. *Desalin. Water Treat.* **2016**, *57*, 17300–17307. [CrossRef]
20. Rambabu, K.; Velu, S. Modified polyethersulfone ultrafiltration membrane for the treatment of tannery wastewater. *Int. J. Environ. Studies* **2016**, *73*, 819–826. [CrossRef]
21. Hasan, K.S.; Visvanathan, C.; Ariyamethee, P.; Chantaraaumporn, S.; Moongkhumklang, P. Vibratory shear enhanced membrane process and its application in starch wastewater recycle. *Songklanakarin J. Sci. Technol.* **2002**, *24*, 899–906.
22. Zouboulis, A.I.; Petala, M.D. Performance of VSEP vibratory membrane filtration system during the treatment of landfill leachates. *Desalination* **2008**, *222*, 165–175. [CrossRef]
23. Petala, M.D.; Zouboulis, A.I. Vibratory shear enhanced processing membrane filtration applied for the removal of natural organic matter from surface waters. *J. Memb. Sci.* **2006**, *269*, 1–14. [CrossRef]
24. AlAkoum, O.; Jaffrin, M.Y.; Ding, L.; Paullier, P.; Vanhoutte, C. An hydrodynamic investigation of microfiltration and ultrafiltration in a vibrating membrane module. *J. Membr. Sci.* **2002**, *197*, 37–52. [CrossRef]
25. AlAkoum, O.; Ding, L.; Chotard-Ghodsnia, R.; Jaffrin, M.Y.; Gesan-Guiziou, G. Casein micelles separation from skimmed milk using a VSEP dynamic filtration module. *Desalination* **2002**, *144*, 325–330. [CrossRef]
26. AlAkoum, O.; Jaffrin, M.Y.; Ding, L.H.; Frappart, M. Treatment of dairy process waters using a vibrating filtration system and NF and RO membranes. *J. Membr. Sci.* **2004**, *235*, 111–122. [CrossRef]
27. Available online: www.vsep.com (accessed on 3 April 2019).
28. Panniza, M.; Cerisola, G. Electrochemical oxidation as a final treatment of synthetic tannery wastewater. *Environ. Sci. Technol.* **2004**, *38*, 5470–5475. [CrossRef]
29. APHA; AWWA; WPCF. *Standard Methods for the Examination of Water and Waste Water*, 19th ed.; American Public Health Association, American Water Works Association, Water Environment Federation: Washington, DC, USA, 1992.
30. Huuhilo, T.; Vaisanen, P.; Nuortila-Jokinen, J.; Nystrom, M. Influence of shear on flux in membrane filtration of integrated pulp and paper mill circulation water. *Desalination* **2001**, *141*, 245–258. [CrossRef]
31. Ulbricht, M. State-of-the-art and perspectives of organic materials for membrane preparation. In *Comprehensive Membrane Science and Engineering*; Drioli, E., Giorno, L., Fontananova, E., Eds.; Elsevier Science: Amsterdam, The Netherlands, 2017; pp. 85–120.
32. Hilal, N.; Ogunbiyi, O.O.; Miles, N.J.; Nigmatullin, R. Methods Employed for Control of Fouling in MF and UF Membranes: A Comprehensive Review. *Sep. Sci. Technol.* **2005**, *40*, 1957–2005. [CrossRef]
33. Bian, R.; Yamamoto, K.; Watanabe, Y. The effect of shear rate on controlling the concentration polarization and membrane fouling. *Desalination* **2000**, *131*, 225–236. [CrossRef]
34. Akamatsu, K.; Ishizaki, K.; Yoshinaga, S.; Nakao, S.-I. Mass transfer coefficient of tubular ultrafiltration membranes under high-flux conditions. *AICHE J.* **2018**, *64*. [CrossRef]
35. Hermia, J. Constant pressure blocking filtration lows: Application to power-low non-newtonian fluids. *Trans. Inst. Chem. Eng.* **1982**, *60*, 183–187.
36. VincentVela, M.C.; Blanco, S.A.; Garcia, J.L.; Rodriguez, E.B. Analysis of membrane pore blocking models applied to the ultrafiltration of PEG. *Sep. Purif. Technol.* **2008**, *62*, 489–498.
37. Hwang, K.J.; Liao, C.Y.; Tung, K.L. Effect of membrane pore size on the particle fouling in membrane filtration. *Desalination* **2008**, *234*, 16–23. [CrossRef]
38. Lim, A.L.; Bai, R. Membrane fouling and cleaning in microfiltration of activated sludge wastewater. *J. Membr. Sci.* **2003**, *216*, 279–290. [CrossRef]
39. Salahi, A.; Abbasi, M.; Mohammadi, T. Permeate flux decline during UF of oily wastewater: Experimental and modelling. *Desalination* **2010**, *251*, 153–160. [CrossRef]
40. Mohammadi, T.; Kazemimoghadam, M.; Saadabadi, M. Modeling of membrane fouling and flux decline in reverse osmosis during separation of oil in water emulsions. *Desalination* **2003**, *157*, 369–375. [CrossRef]
41. Hwang, K.-J.; Lin, T.-T. Effect of morphology of polymeric membrane on the performance of cross-flow microfiltration. *J. Membr. Sci.* **2002**, *199*, 41–52. [CrossRef]

© 2019 by the authors. Licensee MDPI, Basel, Switzerland. This article is an open access article distributed under the terms and conditions of the Creative Commons Attribution (CC BY) license (http://creativecommons.org/licenses/by/4.0/).

Article

Dissolution Testing of Single- and Dual-Component Thyroid Hormone Supplements

Samantha L. Bowerbank, Michelle G. Carlin and John R. Dean *

Department of Applied Sciences, Northumbria University, Ellison Building, Newcastle upon Tyne NE1 8ST, UK; samantha.bowerbank@northumbria.ac.uk (S.L.B.); m.carlin@northumbria.ac.uk (M.G.C.)
* Correspondence: John.Dean@northumbria.ac.uk; Tel.: +44-191-227-3047

Received: 27 November 2018; Accepted: 6 March 2019; Published: 26 March 2019

Abstract: A method for the analysis of thyroid hormones by liquid chromatography-mass spectrometry was used for the dissolution testing of single- and dual-component thyroid hormone supplements via a two-stage biorelevant dissolution procedure. The biorelevant media consisted of fasted-state simulated gastric fluid and fasted state simulated intestinal fluid at 37 °C, and was investigated using an internationally recognized protocol. The dissolution profiles showed consistent solubilization for both single- and dual-component batches at pH 6.5 in the fasted-state simulated intestinal fluid.

Keywords: thyroid; dissolution; liquid chromatography-mass spectrometry

1. Introduction

Thyroid hormones are responsible for the regulation of a variety of metabolic functions, including basal metabolic rate and lipid, glucose and carbohydrate metabolism [1]. This group of compounds contains tyrosine-based compounds including the physiologically active form triiodothyronine (T3) and the prehormone thyroxine (T4). The majority of triiodothyronine is formed enzymatically by the deiodination of thyroxine [2,3] (Figure 1). Hyperthyroidism and hypothyroidism are the two main medical conditions associated with thyroid hormone levels. Hypothyroidism is caused by a depleted level of triiodothyronine, the treatment for which is lifelong thyroid supplement therapy [1,4–7]. Currently, the favoured treatment for hypothyroidism is the administration of levothyroxine sodium salt, with its cost being a considerable factor [8]. However, a number of studies have found that in order to maintain euthyroid levels of both T4 and T3, an excess of levothyroxine sodium salt must be administered [9–11]. This has resulted in products becoming available that contain either T3 alone or a combination of T4 and T3 in a single dose.

Dissolution testing is widely accepted within the pharmaceutical industry as the measure of drug release rate to aid in quality control, formulation and process development [12,13]. Noyes and Whitney, investigating the dissolution of benzoic acid and lead chloride, performed the first reported study into dissolution in 1897 [14]. However, the importance of dissolution testing for pharmaceutical quality control and drug formulation was not established until 70 years later [12,15]. Within the pharmaceutical industry, there are two main categories of dissolution testing performed, biorelevant and quality control [13]. Biorelevant is an abbreviated term for "biologically relevant", and the selected media mimic the fluids found within the stomach (gastric) and intestinal tract [16].

Biorelevant dissolution is a multiple-stage dissolution test designed to model the different in vivo environments as the dosage passes through the gastrointestinal (GI) tract. A biorelevant dissolution method is utilized during early formulation selection and optimization but due to the cost and complexity of the media plus variability of physiological parameters it is replaced by a quality control dissolution method once formulation has been developed [13,15]. A quality control dissolution method consists of one medium that is designed for detecting variations in routine manufacturing or changes during stability testing (e.g., to detect incorrect granulation or compression [13]).

This paper investigates the bioavailability profile, using simulated in vitro gastrointestinal extraction media, of thyroid hormone supplements that contain either a single or a dual combination of the two hormones using biorelevant dissolution media. In addition, the bioavailability profile is compared against the standard United States Pharmacopoeia (USP) quality control test specification of 70% release within 45 min for individual dosage forms of thyroid hormone supplements [17,18].

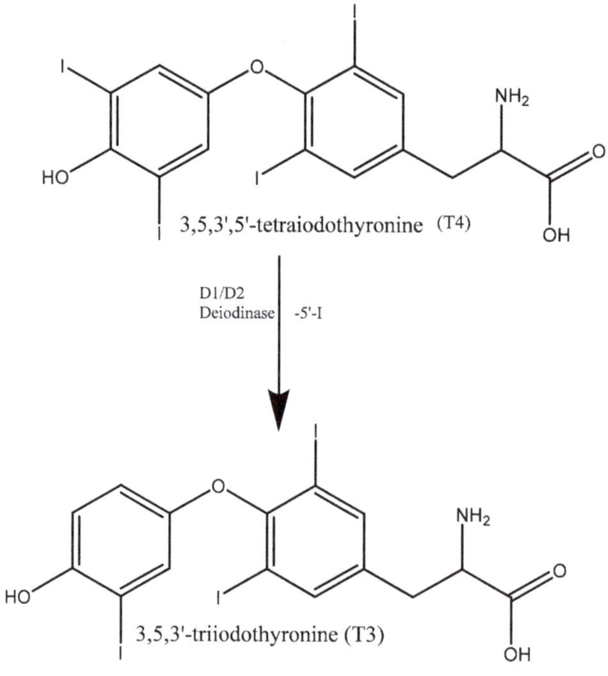

Figure 1. Conversion of T4 to T3 by deiodination.

2. Materials and Methods

2.1. Chemicals and Reagents

Triiodothyronine and thyroxine/triiodothyronine tablets were purchased from RX Cart (Sagunto, Valencia, Spain), specifically Cytomel containing 25 µg of T3. Tiromel containing 25 µg of T3 and Dithyron containing 12.5 µg of T3 and 50 µg of T4 were purchased from http://www.buyt3.co.uk/. Organic LC–MS grade solvents (methanol, acetonitrile, acetic acid and formic acid) were purchased from Fisher Scientific (Loughborough, Leicestershire, UK). Standards of T3 and T4 were purchased from Sigma Aldrich (Poole, Dorset, UK) with a purity of ≥98%. For dissolution media, FaSSIF/FeSSIF/FaSSGF powder was purchased from Biorelevant (London, UK) and additives (sodium hydroxide pellet and monobasic sodium phosphate monohydrate) were purchased from Sigma Aldrich (Poole, Dorset) while sodium chloride, hydrochloric acid and sodium hydroxide were purchased from Fisher Scientific (Loughborough, Leicestershire, UK).

2.2. Instrumentation

LC–MS analysis chromatographic separation was achieved on a reversed phase pentafluorophenyl column (Supelco 2.1 µm F5, 100 × 2.1 mm) from Sigma Aldrich (Poole, Dorset, UK). Thermo

Surveyor LC (Thermo Scientific, Hemel Hempsted, UK) consisted of a quaternary MS pump, vacuum degasser, thermostated autosampler (set to 5 °C) and a thermostated column oven (set to 25 °C). Mass spectrometry was performed using an LTQ XL ion trap mass spectrometer (Thermo Scientific, Hemel Hempsted, UK) equipped with a heated electrospray ionization (HESI) source maintained at 200 °C. The solvent evaporation was aided with auxiliary gas, sheath gas and sweep gas set to an arbitrary flow rate of 15, 60 and 1, respectively. The mass spectrometer was operated in selected reaction monitoring (SRM) MS/MS in negative mode, with collision energies of 28 eV for T4 and 27 eV for T3. The monitored transitions were 776 → 604 and 650 → 633 for T4 and T3, respectively. In SRM MS/MS mode the precursor ion is isolated and subjected to a specified amount of collision energy to induce fragmentation. The MS method is then set to monitor the precursor and a minimum of two stable product ions. Sample aliquots of 10 µL were introduced onto the column at a flow rate of 200 µL/min. The analytes were separated using an isocratic method using water +0.2% formic acid (A) and methanol +0.2% formic acid (B) as the mobile phase.

2.3. Dissolution

Dissolution was carried out using a SOTAX Smart AT7™ dissolution bath from Sotax (Finchley, London). The dissolution bath was set to 37 °C and 250 mL of fasted-state simulated gastric fluid (FaSSGF) was added to each vessel; the rotor speed set to 75 RPM and allowed to equilibrate for at least 1 h. The initial and final temperatures and pH were recorded to ensure consistent temperature and buffer control. Tablet weights were recorded prior to being released into the relevant vessels simultaneously. Samples were taken at the following time points through probe filters and syringe filters: 5, 10, 20 and 30 min. The paddles were stopped and 250 mL of fasted-state simulated intestinal fluid (FaSSIF), added to each vessel and the paddle was resumed. Samples were taken at the following time points through probe filters and syringe filters: 35, 40, 50, 60, 90, 120, 150, 180, 210 and 240 min.

2.4. Preparation of Stock Solutions and Dissolution Media

Stock solutions of each hormone were prepared at a concentration approximately 0.5 mg/mL in methanol, aliquoted into 100 µL aliquots and stored at −20 °C, as recommended by the manufacturer to increase the working life of the standard solutions. A fresh working standard solution of both standards was prepared each week by dilution of stock solutions in mobile phase (70% methanol: 30% water, *v/v*). Calibration standards were prepared daily for each analysis from the working stock solution ranging from 1 to 200 ng/mL for LC–MS/MS. A quality control standard containing both thyroid hormones was also prepared at a concentration of 50 ng/mL.

To prepare the fasted-state small intestinal fluid (FaSSIF) buffer, 0.84 g of sodium hydroxide pellet, 7.90 g of monobasic sodium phosphate monohydrate and 12.38 g of sodium chloride were added to approximately 1.8 L of Type 1 water. The solution was pH adjusted to 6.5 with 1 N sodium hydroxide and made up to 2 L with Type 1 water. To make the final FaSSIF medium, 4.48 g of FaSSIF/FeSSIF/FaSSGF powder was added to approximately 1 L of FaSSIF buffer. The solution was stirred until completely dissolved and made up to 2 L with FaSSIF buffer. The solution was allowed to stand for 2 h prior to use [19]. To prepare the fasted-state gastric fluid (FaSSGF) buffer, 3.2 g of sodium chloride was added to approximately 1.8 L of Type 1 water. The solution was pH adjusted to 1.6 with 1 N hydrochloric acid and made up to 2 L with Type 1 water. To make the final FaSSGF medium, 0.12 g of FaSSIF/FeSSIF/FaSSGF powder was added to approximately 1 L of FaSSGF buffer. The solution was stirred until completely dissolved and made up to 2 L with FaSSGF buffer [19].

2.5. Tablet Analysis

For each batch of tablets, three tablets were weighed and crushed using a pestle and mortar. In triplicate for each tablet, a fifth of the weight was transferred to a 10 mL volumetric flask, made up to volume with water, and sonicated for 20 min. The solutions were allowed to cool, then 10 µL was transferred to a separate 10 mL volumetric flask and made up to volume with water. The solution was then filtered through a 0.22 μm nylon filter and placed in autosampler vials for analysis by LC–MS/MS.

3. Results & Discussion

3.1. Tablet Analysis

Calibration data was generated for both T3 and T4 over the concentration range 0–250 ng/mL, based on 11 data points. Subsequently, the calibration curve obtained gave an r^2 value of 0.9962 and 0.9973 for T3 and T4, respectively (Table 1). The developed LC–MS/MS method was both sensitive and selective for T3 and T4 analysis, with limit of quantitation (LOQ) values of 1.6 and 1.3 ng/mL and limit of detection (LOD) values of 0.2 and 0.8 ng/mL, respectively. Calculations were based on the standard curve method: LOD = (3.3σ)/S and LOQ = (10σ)/S, where σ is the standard deviation and S is the slope of the curve [20]. In addition, a quality control standard (50 ng/mL) was analyzed throughout the experimental duration and given average recoveries of 99.6% and 102.6% for T3 and T4, respectively. Tablet analysis was performed prior to dissolution testing to ensure that the stated dosage was present, as this would be used to indicate 100% of release under dissolution testing. The tablet analysis was consistent with the dosages stated on the packaging for all four batches of thyroid supplement with a % content of >93.3% for all batches and replicate preparations (Table 2). Good inter-batch and inter-tablet precision were also observed with % (relative standard deviation) RSD of <1.8% and <3.4%, respectively.

Table 1. Analytical data for T3 and T4 by LC–MS/MS.

Compound	Calibration Range (ng/mL)	No of Data Points	Linearity	R^2 Value	LOD (ng/mL)	LOQ (ng/mL)
T3	0–250	11	Y = 83.622x − 196.97	0.9962	0.2	1.6
T4	0–250	11	Y = 27.338x + 10.442	0.9973	0.8	1.3

LOD = limit of detection; LOQ = limit of quantitation.

Table 2. Tablet analysis: single- and dual-component thyroid hormone supplements.

Single or Dual Thyroid Hormone Component	Sample	Thyroid Hormone	Replicate	μg per Tablet (n = 3) ± SD	% Content (n = 3) (%RSD)	% Content per Batch (n = 9) (%RSD)
single	Cytomel (batch 1)	T3	1	24.4 ± 0.83	97.6 (3.4)	95.6 (1.8)
			2	23.7 ± 0.27	95.0 (1.2)	
			3	23.6 ± 0.58	94.3 (2.4)	
single	Cytomel (batch 2)	T3	1	23.4 ± 0.46	93.6 (2.0)	94.0 (0.8)
			2	23.7 ± 0.38	94.8 (1.6)	
			3	23.4 ± 0.51	93.5 (2.2)	
single	Tiromel	T3	1	23.3 ± 0.12	93.3 (0.5)	94.4 (1.3)
			2	23.9 ± 0.06	95.7 (0.3)	
			3	23.5 ± 0.19	94.1 (0.8)	
dual	Dithyron	T3	1	12.4 ± 0.08	99.2 (0.6)	96.3 (2.7)
			2	11.8 ± 0.13	94.4 (1.1)	
			3	11.9 ± 0.07	95.2 (0.6)	
	Dithyron	T4	1	49.3 ± 0.21	98.6 (0.4)	98.4 (0.9)
			2	48.7 ± 0.09	97.4 (0.2)	
			3	49.6 ± 0.16	99.2 (0.3)	

3.2. Dissolution

In accordance with the United States Pharmacopeia (USP), the tablets are considered to be dissolved when 75% of the stated dosage has been released [17,18]. The USP method is used as a quality control method, which is normally deployed to identify variations during manufacturing or storage stability, and does not mimic the different physiological conditions of the gastrointestinal tract. The USP-stated method uses alkaline borate buffer (pH 10) and 0.01 N hydrochloric acid containing 0.2% sodium lauryl sulfate for T3 and T4, respectively, with no proposed method for dual-component tablets [20]. Therefore, to ensure consistency and allow a direct comparison between single- and dual-component tablets, biorelevant dissolutions were performed using simulated gastric and intestinal fluids. The results from the dissolution testing using the simulated gastric and intestinal fluids of the tablets for the single- and dual-component thyroid hormones are shown in Figure 2. It is noted that the dissolution for T3 and T4 occurs within approximately 45 min (i.e., 75% dissolution), which is in agreement with the USP method [17,18]. From Figure 2 it can also be seen that total release of T3 and T4 from the tablet formulations was obtained within 120 min.

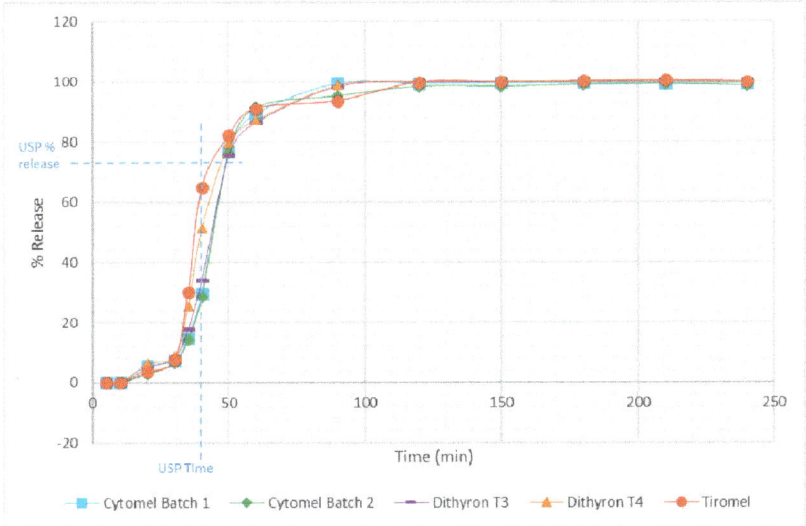

Figure 2. Dissolution testing profile for T3 and T4 from single- and dual-supplement thyroid hormone supplements.

The kinetics of dissolution have been investigated using a first-order rate constant (Figure 3). It was observed that a two-stage dissolution process occurs with crossing points determined by extrapolation of the line of best fit. The initial rate constant corresponds to tablet coating dissolution while the second rate constant is indicative of tablet breakdown. The rate constants (k) were calculated to be between 5.3–6.1 h^{-1} and 0.4–0.8 h^{-1} for coating removal and tablet solubilization, respectively. The rate constant was calculated based on the change in cumulative % drug remaining over time $((y2 - y1)/(x2 - x1))$ with y calculated using the equation of the line.

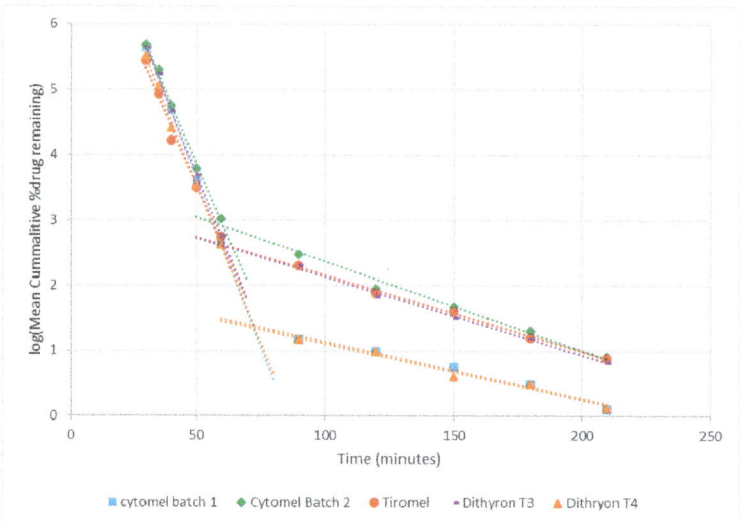

Figure 3. Kinetics of dissolution for T3 and T4 from single- and dual-supplement thyroid hormone supplements.

4. Conclusions

The tablet analysis shows that the thyroid hormone supplements contain T3 and T4 content corresponding to the stated dosage. The dissolution testing profiles and kinetic of dissolution plots show that there is consistent solubilization of the active pharmaceutical ingredient across both single- and dual-component batches for thyroid hormone supplements. Therefore, it is concluded that the use of simulated in vitro gastric intestinal fluids has no influence on dual-component thyroid hormone supplement extraction and recovery. It was also noted, from the dissolution testing profile, that there was minimal solubilization of T3 and T4 in the gastric fluid. However, rapid release of the active compounds was observed within 15 min of the addition of the intestinal fluid.

Author Contributions: Conceptualization, J.R.D., M.G.C. and S.L.B.; methodology, S.L.B.; validation, S.L.B.; formal analysis, S.L.B.; investigation, S.L.B.; resources, S.L.B., J.R.D. and M.G.C.; data curation, S.L.B.; writing—original draft preparation, S.L.B.; writing—review and editing, J.R.D., S.L.B. and M.G.C.; supervision, J.R.D. and M.G.C.; project administration, J.R.D.

Funding: This research received no external funding.

Acknowledgments: We acknowledge GlaxoSmithKline for the donation of the dissolution bath to Northumbria University. We also gratefully acknowledge financial support from Northumbria University.

Conflicts of Interest: The authors declare no conflict of interest. The funders had no role in the design of the study; in the collection, analyses, or interpretation of data; in the writing of the manuscript; or in the decision to publish the results.

References

1. Sinha, R.A.; Singh, B.K.; Yen, P.M. Thyroid hormone regulation of hepatic lipid and carbohydrate metabolism. *Trends Endocrinol. Metab.* **2014**, *25*, 538–545. [CrossRef] [PubMed]
2. Goodman, H.M. *Basic Medical Endocrinology*, 4th ed.; Academic Press: Amsterdam, The Netherlands, 2009; p. 17.
3. Gu, J.; Soldin, O.P.; Soldin, S.J. Simultaneous quantification of free triiodothyronine and free thyroxine by isotope dilution tandem mass spectrometry. *Clin. Biochem.* **2007**, *40*, 1386–1391. [CrossRef] [PubMed]
4. Chaker, L.; Bianco, A.C.; Jonklaas, J.; Peeters, R.P. Hypothyroidism. *Lancet* **2017**. [CrossRef]
5. Dunn, D.; Turner, C. Hypothyroidism in women. *Nurs. Women's Health* **2016**, *20*, 93–98. [CrossRef] [PubMed]

6. Gilbert, J. Hypothyroidism. *Medicine* **2017**, *45*, 506–509. [CrossRef]
7. National Health Service. Hypothyroidism. Available online: https://www.nhs.uk/conditions/underactive-thyroid-hypothyroidism/ (accessed on 24 October 2018).
8. Escobar-Morreale, H.F.; Botella-Carretero, J.I.; Morreale de Escobar, G. Treatment of hypothyroidism with levothyroxine or a combination of levothyroxine plus L-triiodothyronine. *Best Pract. Res. Clin. Endocrinol. Metab.* **2015**, *29*, 57–75. [CrossRef] [PubMed]
9. Escobar-Morreale, H.F.; Obregon, M.J.; Escobar del Rey, F.; Morreale de Escobar, G. Replacement therapy for hypothyroidism with thyroxine alone does not ensure euthyroididm in all tissues, as studied in thyroidectomized rats. *J. Clin. Investig.* **1995**, *96*, 2828–2838. [CrossRef] [PubMed]
10. Escobar-Morreale, H.F.; Obregon, M.J.; Hernandez, A.; Escobar del Rey, F.; Morreale de Escobar, G. Regulation of iodothyronine deiodinase activity as studied in thyroidectomized rats infused with thyroxine or triiodothyronine. *Endocrinology* **1997**, *138*, 2559–2568. [CrossRef] [PubMed]
11. Escobar-Morreale, H.F.; Obregon, M.J.; del Escobar Rey, F.; Morreale de Escobar, G. Tissue-specific patterns of changes in 3,5,3-triiodo-L-thyronine concentrations. *Biochimie* **1999**, *81*, 453–462. [CrossRef]
12. Dokoumetzidis, A.; Macheras, P. A century of dissolution research: From Noyes and Whitney to the biopharmaceutics classification system. *Int. J. Pharm.* **2006**, *321*, 1–11. [CrossRef] [PubMed]
13. Grady, H.; Elder, D.; Webster, G.K.; Mao, Y.; Lin, Y.; Flanagan, T.; Mann, J.; Blanchard, A.; Cohen, M.J.; Lin, J.; et al. Industry's view on using quality control, biorelevant, and clinically relevant dissolution tests for pharmaceutical development, registration, and commercialization. *J. Pharm. Sci.* **2018**, *107*, 34–41. [CrossRef] [PubMed]
14. Noyes, A.A.; Whitney, W.R. The rate of solution of solid substances in their own solutions. *J. Am. Chem. Soc.* **1897**, *19*, 930–934. [CrossRef]
15. Dressman, J.B.; Kramer, J. *Pharmaceutical Dissolution Testing*, 1st ed.; Taylor & Francis: Boca Raton, FL, USA, 2005; pp. 4–15.
16. Leigh, M.; Kloefer, B.; Schaich, M. Comparison of the solubility and dissolution of drugs in fasted-state biorelevant media (FaSSIF and FaSSIF-V2). *Diss. Technol.* **2013**, *20*, 44–50. [CrossRef]
17. United States Pharmacopeia. USP36 NF31. In *Official Monographs/Levothyroxine*; United States Pharmacopeia: Rockville, MD, USA, 2013; pp. 4109–4110.
18. United States Pharmacopeia. USP36 NF31. In *Official Monographs/Liothyronine*; United States Pharmacopeia: Rockville, MD, USA, 2013; pp. 4121–4123.
19. Biorelevant. Biorelevant Instructions V1.1. Available online: https://biorelevant.com/ (accessed on 30 October 2018).
20. ICH Guideline Validation of Analytical Procedures—Test and Methodology. Available online: http://www.ich.org/products/guidelines/quality/quality-single/article/validation-of-analytical-procedures-text-and-methodology.html (accessed on 11 February 2019).

© 2019 by the authors. Licensee MDPI, Basel, Switzerland. This article is an open access article distributed under the terms and conditions of the Creative Commons Attribution (CC BY) license (http://creativecommons.org/licenses/by/4.0/).

Article

Estimating Diphenylamine in Gunshot Residues from a New Tool for Identifying both Inorganic and Organic Residues in the Same Sample

Ana Isabel Argente-García [1], Lusine Hakobyan [1], Carmen Guillem [2] and Pilar Campíns-Falcó [1,*]

[1] MINTOTA Research Group, Departament de Química Analítica, Facultat de Química, Universitat de València, Dr. Moliner 50, Burjassot, 46100 Valencia, Spain; a.isabel.argente@uv.es (A.I.A.-G.); snkhchyan@yahoo.com (L.H.)
[2] Institut de Ciència dels Materials (ICMUV), Universitat de València, P.O. Box 22085, 46071 Valencia, Spain; carmen.guillem@uv.es
* Correspondence: pilar.campins@uv.es; Tel.: +34-96-354-3002

Received: 10 January 2019; Accepted: 26 February 2019; Published: 19 March 2019

Abstract: A method involving the collection and determination of organic and inorganic gunshot residues on hands using on-line in-tube solid-phase microextraction (IT-SPME) coupled to miniaturized capillary liquid chromatography with diode array detection (CapLC-DAD) and scanning electron microscopy coupled to energy dispersion X-ray (SEM-EDX), respectively, for quantifying both residues was developed. The best extraction efficiency for diphenylamine (DPA) as the main target among organic residues was achieved by using a dry cotton swab followed by vortex-assisted extraction with water, which permits preservation of inorganic residues. Factors such as the nature and length of the IT-SPME extractive phase and volume of the sample processed were investigated and optimized to achieve high sensitivity: 90 cm of TRB-35 (35% diphenyl, 65% polydimethylsiloxane) capillary column and 1.8 mL of the processed sample were selected for the IT-SPME. Satisfactory limit of detection of the method for analysis of DPA deposited on shooters' hands (0.3 ng) and precision (intra-day relative standard deviation, 9%) were obtained. The utility of the described approach was tested by analyzing several samples of shooters' hands. Diphenylamine was found in 81% of the samples analyzed. Inorganic gunshot residues analyzed by SEM-EDX were also studied in cotton swab and lift tape kit samplers. Optical microscopy was used to see the inorganic gunshot residues in the cotton swab samplers. The lift tape kits provided lesser sensitivity for DPA than dry cotton swabs—around fourteen times. The possibility of environmental and occupational sources could be eliminated when DPA was found together with inorganic residues. Then, the presence of inorganic and organic residues in a given sample could be used as evidence in judicial proceedings in the forensic field.

Keywords: diphenylamine; gunshot residues; hands; dry cotton swab; in-tube solid-phase extraction; capillary liquid chromatography; SEM-EDX

1. Introduction

Chemical and physical evidence such as gunshot residues (GSRs) from firearms discharge may provide valuable forensic information [1,2]. Gunshot residues are organic and inorganic components in nature, which can be deposited on a shooter's body, mainly onto the index fingers and thumbs of the hands, after discharging a firearm [3]. A suspect can be successfully identified if GSRs are reliably analyzed. Thus, the detection of these compounds plays an important role in the field of forensic science. Inorganic gunshot residues (IGSRs) are usually spherical particles mainly composed of Ba, Pb, and Sb [4]. Other elements such as Ca, Al, Cu, Fe, Zn, Ni, Si, and K can also be found [5], although they are more prevalent in the environment than Pb, Ba, and Sb [6]. The size of these particles is usually from 0.5 μm to 10 μm, although sizes up to 100 μm have also been reported [7]. The presence of these metallic particles has been traditionally confirmed by scanning electron microscopy coupled to the energy dispersion X-ray (SEM-EDX) technique due to its non-destructive capability to perform both morphological and elemental analyses [8,9]. However, the analysis of IGSRs has its limitations. False positive results can be produced from inorganic particles derived from environmental and occupational sources [10–13], which is a problem when considering IGSRs as evidence in judicial proceedings in the forensic field. The analysis of organic gunshot residues (OGSRs) in the same sample could provide complementary information that could strengthen the probative value of a forensic sample. Organic components originate mostly from the propellant, and their composition depends on the commercial brand and ammunition type.

An important component of gun propellants is diphenylamine (DPA), which is used as a stabilizer in order to prevent the decomposition of explosive products like nitrocellulose and nitroglycerine, both of them present in many smokeless powders used as propellants [14]. Thus, this stabilizer may remain on a shooter's hands, and it may be used as an indicator of gunshot residues [15]. Diphenylamine detection could provide valuable evidence of firearm discharge for the identification of suspects in firearm-related crimes.

The low amount of DPA remaining on a shooter's hands requires highly-sensitive analytical techniques for its detection. In order to improve the sensitivity, many methods include off-line sample treatment, which involves time-consuming and tedious steps. Table 1 presents several methods used for extraction and determination of DPA that remains on the hands. The main drawback of the reported methods is the low detection limit required, taking into account the sampling and extraction process, time of analysis, and greenness of the procedure.

Table 1. Comparison of reported methods for determining diphenylamine (DPA) on a shooter's hands. The method proposed in this work was also included for comparison (On-line in-tube solid phase microextraction coupled to capillary liquid chromatography with diode array detection (IT-SPME-CapLC-DAD).

Technique/Limit of Detection (LOD)	DPA Extraction	DPA Amount on Hands	Mobile Phase; Flow; Injection Volume	Organic Solvents	Ref.
High-performance liquid chromatography-tandem mass spectrometry/ 0.3 ng/mL (solution)	DPA was extracted with cotton swab soaked with acetone, which was evaporated and DPA was dissolved in 0.1 mL methanol.	<Limit of quantitation (LOQ)	Methanol-water (90:10); 800 μL/min; 10 μL	Methanol and acetone as extractive solvents and mobile phase	[15]
Gas chromatography-mass spectrometry/ 3 ng	DPA was extracted with cotton swab moistened in water, the swab was heated and capillary microextraction made	≈1 ng < LOQ	-	Water as extractive solvent	[4]
Tandem Mass Spectrometry/ 1 ng/mL (solution)	Cotton swab soaked with methanol to extract DPA from the hand and dilution to 1 mL of methanol	Not studied	0.1 mL/min; 20 μL	Methanol as extractive solvent	[16]
Mass spectrometry/-	Dabbing an adhesive coated aluminum stub over the hands	Not detected	4 μL/min	Water:methanol 0.1% formic acid as solvent spray	[17]

Table 1. *Cont.*

Technique/Limit of Detection (LOD)	DPA Extraction	DPA Amount on Hands	Mobile Phase; Flow; Injection Volume	Organic Solvents	Ref.
Liquid chromatography-tandem mass spectrometry/ 34,000 ng	Cotton swab moistened with isopropyl alcohol:water, 75:25, which was introduced in a tube with 3.2 mL of the mixture and centrifuged. The aliquot was diluted five times with deionized water. SPEC C18 cartridges were conditioned with 250 µL of isopropyl alcohol and deionized water. 5000 µL of aqueous samples were loaded. The sorbent was rinsed with 250 µL of deionized water and dried. The analytes were eluted in acetonitrile:water:methyl alcohol, 80:10:10; 200 µL	0.29–83 nmol/L	Acetonitrile: methanol: water, acidified by 0.1% of formic acid; 200 µL/min; 20 µL	Isopropylalcohol as extraction solvent, methanol and acetonitrile for mobile phase	[18]
Capillary electrophoresis/ 2387 ng/mL (solution)	Hands were swabbed by a cotton swab embedded in a solvent. The analyte was recuperated by sonication into 2 mL of solvent. Liquid extraction was carried out with 2 mL of ethyl acetate and 50 µL of ethylene glycol; the solvent was evaporated under dry nitrogen. The residues were reconstituted with diaminocyclohexane tetraacetic acid, and borate	Not detected	-	Diaminocyclohexane tetraacetic acid and sodium dodecyl sulfate as sampling solvents	[5]
IT-SPMS-CapLC-DAD/ 0.15 ng/mL (solution) 0.3 ng by cotton swab	DPA was extracted from hands by cotton swab and then DPA was extracted to 2 mL of water under vortex conditions (20 s)	<LOD-16.5 ng	Acetonitrile: water gradient; 10 µL/min; 72 µL	Water as extractive solvent. Acetonitrile as mobile phase	This work

On-line sample pre-treatment has become an interesting alternative as green analytical chemistry indicates. In this context, our research group has successfully applied in-tube solid-phase microextraction (IT-SPME) in the analysis of a variety of analytes and matrices [19,20]. In-tube solid-phase microextraction typically uses a capillary column internally coated with extractive phase, which can be different in nature in function of the physical-chemical properties of the analytes [19,20], in order to extract, concentrate, and clean-up the sample. When IT-SPME is coupled to a miniaturized liquid chromatograph, important improvements in terms of sensitivity, selectivity, automation, and waste minimization can be achieved. Although mass spectrometry (MS) coupled to gas chromatography (GC) or liquid chromatography (LC) offers suitable sensitivity, the chromatographic techniques can present issues. Thermal degradation of DPA can occur by GC and the wide range of polarities of compounds present in GSRs can limit the LC. Some methods have also successfully identified DPA using several MS techniques without any chromatographic system such as tandem mass spectrometry (MS–MS) [16], desorption electrospray ionization-mass spectrometry (DESI-MS) [17], nanoelectrospray ionization mass spectrometry (nESI-MS) [21], and ion mobility spectrometry (IMS) [22]. However, IT-SPME coupled to capillary liquid chromatography (CapLC) contributes to increase the sensitivity and sample clean-up in an on-line way. Additionally, the miniaturization of the LC technique (i.e., low column dimensions, low flow rates, low amount of wastes) contributes also to achieve improved sensitivity, which can permit the use of diode array UV-detectors (DADs), which cost less than an MS detector.

In the present work, a shooters' hands sampling was carried out using dry cotton swabs followed by short vortex-assisted extraction of DPA from cotton with water. Additionally, on-line IT-SPME-CapLC-DAD was employed, for the first time to our knowledge, for the DPA determination. Other samplers were also studied, but their extraction capacities were lower than that achieved by a dry cotton swab. Several parameters such as capillary length and coating, as well as extraction conditions, were optimized for the on-line system. On the basis of the results obtained, a new approach is proposed for the detection of DPA from shooters' hands, which integrates simple, rapid, and green extraction followed by on-line clean-up and preconcentration of samples. The method permits to carry out the analysis of IGSRs by SEM-EDX after the DPA extraction, in order to confirm the presence of inorganic gunshot residues on shooters' hands as well. Optical microscopy can be used for identifying particles with a spherical shape and size up to 20 μm in a cotton swab due to the presence of gunpowder particles, and it was proved that SEM-EDX can be applied after extracting DPA from the swab. The other aim of this work was to examine the morphology and elemental composition and distribution of GSR particles collected with the lift tape kits, the typical police collector, which provided lesser sensitivity in the DPA analysis (around fourteen times less). The possibility of environmental and occupational sources could be eliminated when DPA was found together with IGSRs. Both analyses can be used as evidence in judicial proceedings in the forensic field [23].

2. Materials and Methods

2.1. Materials

All the reagents were of analytical grade. Acetonitrile (ACN) HPLC grade was supplied by Prolabo (Fontenay-sous-Bois, France). Ethanol, acetone, and DPA were purchased from Scharlau (Barcelona, Spain). Stock standard solution of DPA (10 μg/mL) was prepared by dissolving an adequate amount of DPA in acetonitrile. Working solutions of this compound were prepared by dilution of the stock solution with water. Ultrapure water was obtained from a Nanopure II system (Sybron, Barnstead, UK). All solutions were stored in the dark at 4 °C.

Cotton swabs (100% cotton; 0.03 g of the amount of cotton on each tip) from a local market, double-sided carbon adhesive tape (8 mm wide × 0.16 mm thick × 1 cm long; Ted Pella Inc. Redding, CA, US), and tape lift kits (Adhesive Lifts GRA 200, Sirchie Finger Print Laboratories, Youngsville, NC, USA) were employed as sample collectors. Polydimethylsiloxane (PDMS) Sylgard® 184 Silicone Elastomer Kit containing Sylgard® 184 silicone elastomer base and Sylgard® 184 silicone elastomer curing agent, provided by Dow Corning (Midland, MI, USA) and tetraethyl orthosilicate (TEOS) purchased from Sigma–Aldrich (St. Louis, MO, USA), PDMS, and TEOS were used to prepare several samplers. Polydimethylsiloxane base was mixed with TEOS under vigorous magnetic stirring for 10 min at room temperature. Then, a PDMS curing agent was added with a weight ratio of 1:10 to the PDMS base under magnetic stirring for 10 min at room temperature. Finally, 0.02 g of that blend was deposited on well-plates, and then was cured at 30 °C for hours or a day, depending on the film composition (as TEOS increases, curing time increases too). Several weight ratios of PDMS/TEOS were tested (100/0, 50/50, 30/70). The thickness of the film was 1 mm and the diameter was 15 mm.

2.2. Apparatus and Chromatographic Conditions

The capillary chromatographic system used consisted of a capillary liquid chromatography pump (Agilent 1100 Series, Waldbronn, Germany), a high-pressure six-port valve (7725 Reodhyne,

Rohnert Park, CA, USA), an on-line degasser, and an UV-Vis photodiode array detector (Agilent, 1260 Series) equipped with an 80-nL flow cell. The detector was linked to a data system (Agilent, HPLC ChemStation) for data acquisition and calculation. The absorption spectra were recorded between 190 and 400 nm and the chromatograms were monitored at 280 nm. A Zorbax SB-C18 capillary analytical column (150 mm × 0.5 mm i.d., 5 µm particle diameter) was employed for the chromatographic separation (Agilent, Waldbronn, Germany). The mobile phase used was a mixture of acetonitrile:water in gradient elution mode: the initial acetonitrile content was 70% during 1 min, increased to 100% until 12 min, and maintained at 16 min, and then from 16 min to 20 min at 70% acetonitrile. The mobile phase flow rate was 10 µL min^{-1}. All solutions were filtered with 0.45-µm nylon membranes (Teknokroma, Barcelona, Spain) before use.

An ultrasonic bath (300 W, 40 kHz, Sonitech, Guarnizo, Spain) and a ZX3 vortex mixer (40 Hz) from VELP Scientifica (Usmate Velate, Italy) were employed for the lixiviation of the DPA from the sample collectors. An optical microscope (ECLIPSE E200LED MV Series, Nikon Corporation, Tokyo, Japan) under bright-field illumination and using a 10× objective was used to see the collection of inorganic particles on the cotton swab. Nis Elements 4.20.02 software (Nikon Corporation) was used for acquiring the images. In order to test the presence and morphology of IGSRs, scanning electron microscopy (SEM) images were obtained with Hitachi S-4800 FEG (Tokyo, Japan) and Philips XL30 operating at 20 Kv for tape lift kit and cotton swab samples. Au/Pd coating was required. Elemental analysis was performed by an EDX analysis system incorporated into the microscope.

2.3. IT-SPME Procedure

The setup used in this work corresponded to that developed for in-valve IT-SPME [19,20]. The stainless-steel injection loop of a six-port injection valve was replaced with an extractive capillary. Several gas chromatography capillary columns (0.32 mm i.d.) were tested as extractive capillaries. The columns used were TRB-5, TRB-20, TRB-35, TRB-50 (Teknokroma, Barcelona, Spain) and Zebron ZB-WAXplus (Phenomenex, Torrence, CA, USA). For coating details, see Table 2. Segments from 30 to 90 cm of these columns were directly tested for IT-SPME. Capillary connections to the valve were facilitated by the use of 2.5-cm sleeves of 1/16 in polyether ether ketone (PEEK) tubing; 1/16 in PEEK nuts and ferrules were used to complete the connections. In load valve position, 1800 µL of sample was manually passed through the capillary column by means of a 1000-µL precision syringe. A clean-up step was also carried out by processing 120 µL of ultrapure water after the sample loading. Finally, when the valve was manually rotated to the injection position, the analyte was desorbed in dynamic mode from the coating of the extractive capillary and transferred to the analytical column by the mobile phase. The valve was maintained in this position until the end of the chromatogram.

Table 2. Characteristics of capillary columns employed during the in-tube solid-phase microextraction (IT-SPME).

Extractive Capillary	Coating	Coating Thickness (µm)
TRB-5	5% diphenyl-95% polydimethylsiloxane	3
TRB-20	20% diphenyl-80% polydimethylsiloxane	3
TRB-35	35% diphenyl-65% polydimethylsiloxane	3
TRB-50	50% diphenyl-50% polydimethylsiloxane	3
Zebron ZB-WAXplus	polyethylene glycol	1

2.4. Shooting and Collection of GSRs from Hands

Test shots were carried out by police officers in an indoor range at Police Headquarters of Valencian Community (Valencia, Spain) under typical shooting practice conditions. Personal information was not recorded. The shots were fired with 9-mm Heckler & Koch pistols, model USP Compact (Oberndorf/Neckar, Germany), which is the most commonly used firearms among police forces in Spain. Each volunteer police officer fired a total number of 25 shots (regulatory number of shots). Only one of these police officers fired 12 shots because his pistol jammed. In order to avoid contamination, each police officer fired with his own firearm and did not touch other surfaces with their hands during the analysis. Gunshot residue samples were collected from the shooters' hands immediately after discharging the firearm. Sampled zones of the hands are shown in Figure 1. For each police officer, both hands, right and left (palm and back), were sampled after shooting. Two techniques for GSR collection from hands were carried out: swabbing and tape lifting. Swabbing was performed by scrubbing the hand with one of the tips of a cotton swab, which was stored in a 5-mL glass vial with a fitted cap to prevent contamination from other compounds in the air. Note that cotton swabs were not moistened in any solvent before sampling. The tape lift kit consisted of a metal stub equipped with a carbon adhesive tape inserted in a plastic vial with a tightly fitted cap. For the sampling, the metal stub was passed over the surface of the hand and then was returned to the vial. Once all the collected samples were placed back into their vials and capped, they were transported to the lab and were stored at room temperature awaiting analysis. A total of 11 shooters were sampled by swabbing and the other five shooters were sampled by tape lifting, which consisted of a total of 21 swab samples and six tape samples (see Analysis of Samples in the Results and Discussion section for identifying the samples). Additional swab samples from each volunteer police officer before test shots were also analyzed as blanks (hands were not previously washed).

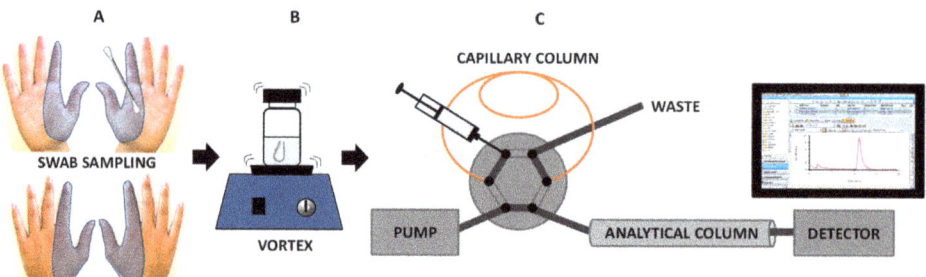

Figure 1. Schematic diagram of the steps for DPA analysis: (**A**) web and palm of the hand sampling (zone sampling in blue), (**B**) vortex-assisted extraction, and (**C**) IT-SPME-capillary liquid chromatography (CapLC) system.

2.5. Sample Treatment for DPA Analysis

Several solvents (water, acetone, ethanol), samplers (cotton swabs, carbon-based tapes, PDMS-TEOS based samplers), extraction techniques (non-assisted, ultrasound-assisted, and vortex-assisted extraction) and time extraction (up to 20 min) were tested in order to find the proper sampling procedure. Three µL of 10 µg/mL in 2 mL of water with different (A) extraction modalities and (B) sample collectors were assayed by IT-SPME-CapLC-DAD. Each sample was analyzed in triplicate and all assays were carried out at ambient temperature.

In order to obtain the solid DPA from standard solutions, a volume (3 µL) of DPA solution (10 µg/mL) in acetonitrile (ACN) was deposited on a glass slide. Then, the solvent was evaporated to dryness at room temperature and solid DPA was collected carefully by scrubbing the glass slide with a cotton swab. After sample collection, the tip of the cotton swab was placed into a storage vial containing 2 mL of water, so that the cotton was completely wetted. Diphenylamine was extracted from the swab under vortex condition for 20 s at ambient temperature. Next, the swab was used in the analysis of inorganic residues, and 1800 µL of the solution was loaded into the IT-SPME capillary of the LC system. The same procedure was used for the other samplers assayed. The complete procedure for the DPA analysis is shown in Figure 1.

2.6. IGSRs Analysis by SEM/EDX and Optical Microscopy

In order to confirm the presence of GSRs in cotton swabs, the gunpowder grains were visually and microscopically identified before chromatographic analysis. For the IGSR particle analysis from the tape lift kits and from cotton swabs, SEM images, EDX spectrums, and X-ray maps were carried out. For cotton swab samplers, besides metallization with Au/Pd coating, silver lac was used for painting the sample. Magnification varied between 50 and 500× according to the particle size. Once the particle was located, an elemental analysis was carried out to determine the major components of the particle. The size, shape, and morphology of the particles were also recorded.

3. Results and Discussion

3.1. Optimization of the IT-SPME and Chromatographic Conditions

Experiments were performed in order to optimize the DPA extraction by IT-SPME, as well as the subsequent chromatographic analysis. Initially, two mobile-phase compositions in isocratic elution were tested, 60:40 and 70:30 ACN: water (v/v). As can be seen in Figure 2A, both compositions were adequate to desorb DPA from the IT-SPME extractive capillary. However, a decrease in retention time and narrower peaks were achieved with the increase of ACN and flow rate of the mobile phase. A gradient elution program (See Section 2.2 for optimum conditions) with 100% of ACN during 4 min was employed as cleaning solvent.

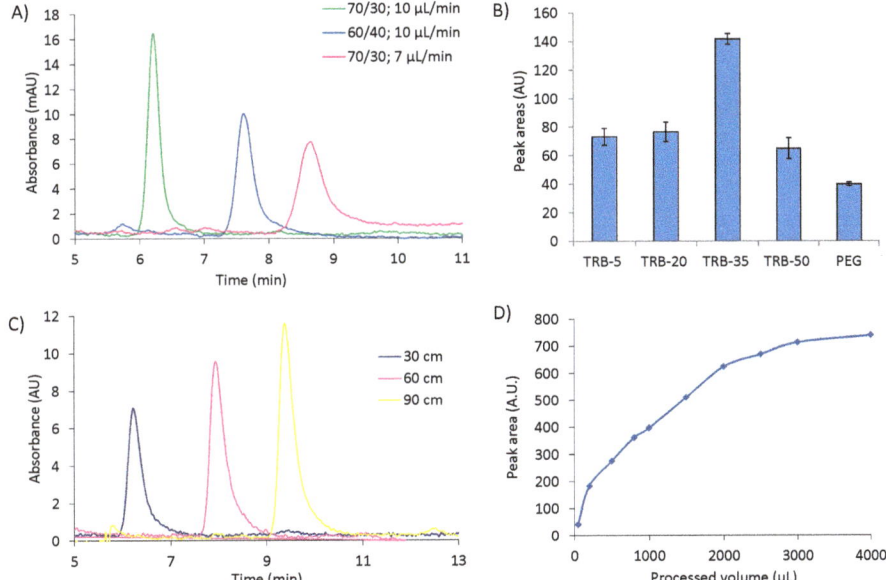

Figure 2. Effect of (**A**) acetonitrile percentage and flow rate of mobile phase (800 µL at 7 ng/mL of DPA, TRB-35, 30 cm); (**B**) nature of the IT-SPME phase (800 µL at 5 ng/mL of DPA, capillary length 30 cm, optimum mobile phase); (**C**) capillary length (800 µL at 5 ng/mL of DPA, TRB-35, optimum mobile phase); and (**D**) sample volume processed (5 ng/mL of DPA, TRB-35, capillary length 90 cm, optimum mobile phase) in IT-SPME versus peak area of DPA. For more details, see the main text.

In-tube solid-phase microextraction was performed using a capillary column as the loop of the injection valve. The analytes were extracted during sample loading and were transferred to the analytical column with the mobile phase by changing the valve position. This configuration was advantageous in order to achieve suitable limit of detection (LOD) for detecting DPA deposited on shooters' hands. Herein, several assays were carried out to optimize the extraction step. The nature of the extractive phase, the length of the capillary column, and the volume of the sample processed were evaluated. Five phases for IT-SPME were assayed: 5, 20, 35, 50% diphenyl–95, 80, 65, 50% polydimethylsiloxanes, respectively, and 100% polyethylene glycol (PEG) (See Table 2). Figure 2B compares the analytical response (mean peak area) for DPA (5 ng/mL) with the different capillaries (30-cm length) when the volume of standard processed was 800 µL. As can be seen, the TRB-35 phase provided higher analytical responses for DPA. This suggests that the higher percentage of diphenyl groups in the extractive phase led to an increase in analytical response. It can be deduced that extraction involves π–π interactions with DPA, whose structure possesses two aromatic rings. However, TRB-50 provided a decrease on the peak area, and this effect was attributed to the increment on the polarity of the extractive phase, and so the affinity towards the DPA decreased (log K_{ow} = 3.5). The same effect may occur by PEG capillary due to its higher polarity. Thus, the TRB-35 capillary column was selected as the best extractive phase for further experiments.

The effect of the capillary length on the analytical response (peak area) was also studied by processing 800 µL of working solution of DPA (5 ng/mL) with TRB-35 capillaries of 30, 60, and 90 cm. Figure 2C shows the increment of the analytical response with the length of the capillary, thus, the amount of analyte extracted also increased. The peak area for DPA improved 40% and 47% with the capillary columns of 60 and 90 cm, respectively, compared with the capillary of 30 cm. Capillaries longer than 90 cm did not improve the analytical response. The TRB-35 of 90 cm was chosen as the optimal capillary column length.

Sample volumes processed up to 4 mL at 5 ng/mL of DPA solution were studied. The results obtained are depicted in Figure 2D. As can be seen, a remarkable increase of analytical response (peak area) with the increase of the sample volume was observed up to 2 mL. The signal increased very slightly from 2 to 4 mL, and 2 mL was chosen as the optimum sample volume for further experiments. However, it was found that the swabs used in the present study absorbed about 125 µL of contact solution. According to this observation, further experiments were carried out by processing 1800 µL remaining in the vial.

The extraction efficiencies of the proposed methodology were estimated by comparing the amount of analyte extracted, which is the amount of the analyte transferred to the analytical column, with the total amount of analyte passed through the extraction capillary. The amount of analyte extracted was established from the peak areas in the resulting chromatograms and from the calibration equations constructed through the direct injection of 72 µL of analyte standard solutions of different concentrations. This volume is the inner volume of the TRB-35 capillary of 90 cm used for IT-SPME. The absolute extraction efficiency obtained was 7% which is in accordance with those reported for this technique [19,20]. Although low extraction efficiencies (absolute recoveries) were achieved by IT-SMPE, the analytical responses were improved significantly owing to the large volumes of sample that can be processed through the capillary column. In addition, a clean-up step was tested after sample loading by introducing 120 µL of nanopure water before changing the valve to the inject position. Significant loss of analyte was not observed; thus, clean-up was applied in order to remove fibers or compounds from cotton which could remain inside the capillary column. It was also tested to filter the solutions of DPA extracted from cotton swab through 0.45-µm nylon membranes. Nevertheless, the analyte was retained on the nylon filter. Hence, samples were not filtered before injection.

3.2. Optimization of DPA Extraction from Hands

3.2.1. DPA Extraction from Collector

The first step considered to optimize DPA extraction was to find an appropriate extraction procedure for DPA from the collector. For this aim, non-assisted, ultrasound-assisted, and vortex-assisted extraction of the analyte from a cotton swab sampler were tested. Three µL of 10 µg/mL DPA working solution (prepared in ACN to favor evaporation) was spread on a glass slide. After it was air evaporated to dryness, a dry cotton swab was used to collect DPA from the slide. The tip of the cotton swab was introduced into a vial containing 2 mL of water under the three abovementioned extraction modes for 5 min (See Section 2.6 for more details). Vortex-assisted extraction offered the best results in terms of both analytical response (peak area) and relative standard deviations (RSDs), as can be seen in Figure 3A. For evaluating the extraction efficiencies, the peak area ratios between non-assisted and assisted extractions were calculated, ratios of 2 and 5 were obtained for ultrasound and vortex, respectively. Moreover, the results provided a satisfactory RSD of 9% for extraction by vortex but not by ultrasounds with 32% of RSD. From these results, we concluded that the best extraction of DPA from the sampler was vortex-assisted extraction.

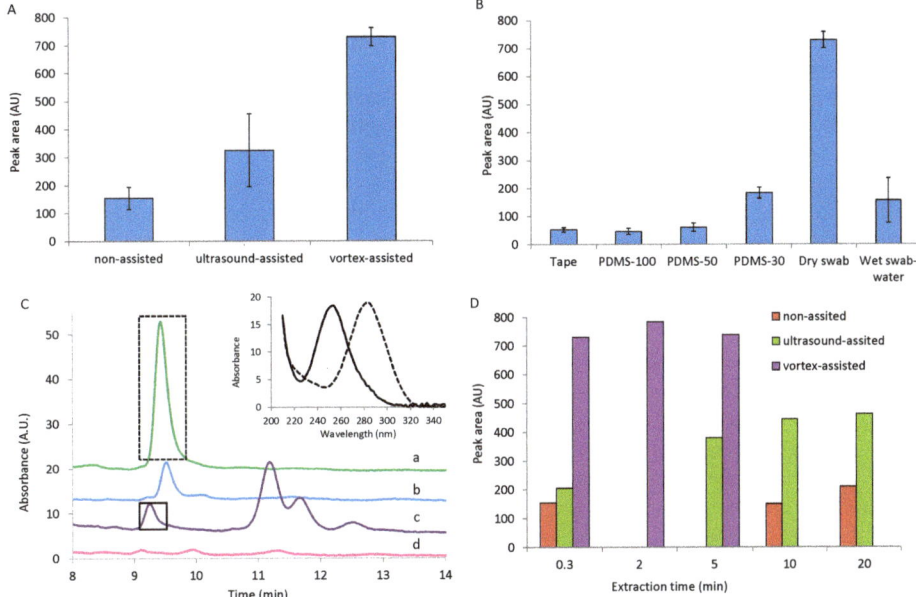

Figure 3. Comparison of peak areas obtained for standard solution (3 µL of 10 µg/mL in 2 mL of water, 15 ng/mL) with different (**A**) extraction modalities with dry swab samplers, (**B**) sample collectors, and (**C**) solvents to wet cotton swabs (at 10 ng/mL): dry (a), water (b), acetone (c), and ethanol (d), together with normalized spectra (inset) of DPA (black dashed line) and unknown compounds (black solid line), and (**D**) extraction time with dry swab samplers. For other experiment details, see main text.

3.2.2. DPA Collecting

Several sampling tools were tested for DPA collection from shooters' hands. In a first attempt, 3 µL of 10 µg/mL DPA working solution was dropped on a glass slide. After it was air evaporated to dryness, solid DPA was collected by several sampling tools: adhesive tape lifts; PDMS-based devices at several PDMS: TEOS proportions (100:0, 50:50, and 70:30); and dry cotton swabs and wet cotton swabs with non-skin-toxic solvents such as water, acetone, and ethanol. According to the 24.

European Chemicals Agency (ECHA) database [24], methanol and acetonitrile were not used due to their harmfulness and toxicity in contact with the skin, respectively. After, the samplers were in contact with 2 mL of water under vortex conditions. Figure 3B compares the mean peak areas of DPA extracted from slides and their RSDs to determine the suitability of the several sampling devices tested. The dry cotton swab achieved the highest analytical response with suitable precision. The adhesive tape lift, which was used in the tape lift kits, showed an analytical response about 14 times lower than the dry cotton swabs. Similar loss of peak area was observed with the pure PDMS-based device. However, increases of analytical response were achieved when the TEOS proportion increased in the composition device. In the case of the PDMS: TEOS (30:70) device, the analytical response was improved by four times, compared with the response with the pure PDMS device. This effect can be attributed to the increment of the device hydrophilicity as a function of the TEOS amount, suggesting the improvement of analyte extraction from device to the aqueous solution. When the cotton swab was wet with water and ethanol, the analytical response decreased 80% and 97%, respectively, compared with the response obtained by a dry swab. It could be due to the wet swab spreading the analyte on the slide surface instead of collecting it; RSD > 30% were obtained indicating the difficulty in controlling the analyte collection. When acetone was used as the extractive solvent, DPA was not detected but a small chromatographic peak at a retention time slightly lower than that of the analyte was observed (Figure 3C). As can be confirmed by the spectra depicted in the Figure 3C inset, this peak could be differentiated from the analyte peak by retention time and spectrum, and it could correspond to some compound from the cotton swab. From these results, a dry cotton swab was chosen as the best sampling collector of DPA from shooters' hands for further work.

Peak areas of DPA were obtained for different extraction times under vortex-assisted extraction of the dry cotton swab sampler: 20 s, 2 and 5 min, as can be seen in Figure 3D and non-significant differences on peak areas were observed. Worse results were achieved with non-assisted and ultrasound-assisted extractions even under higher extraction times. Therefore, 20 s as extraction time was selected by using vortex to extract DPA from hands to suitable level in a short time frame.

3.2.3. Effect of Extraction Solvent on the DPA Extraction from the Sampler

The capacity of three solvents to remove the DPA residues from cotton swabs was investigated: acetonitrile, ethanol, and water. Mixtures of 90:10 water, ACN and water, and ethanol and 100% water were tested. Fifty-one percent and 85% decreases in peak area were observed when ethanol and ACN, respectively, were present in the extraction solvent (See Figure 4). This suggests that the analyte was probably non-retained on the IT-SPME capillary column. Moreover, high peaks were observed at a retention time slightly lower than that corresponding to the analyte. These peaks were not detected when the analysis was carried out in solution, suggesting they were due to compounds extracted from cotton. Note that ethanol was the solvent which extracted more interfering compounds. However, water offered the best results in terms of extraction and reduced interferences, as well as it is a greener solvent. Hence, water was chosen as optimum extraction solvent.

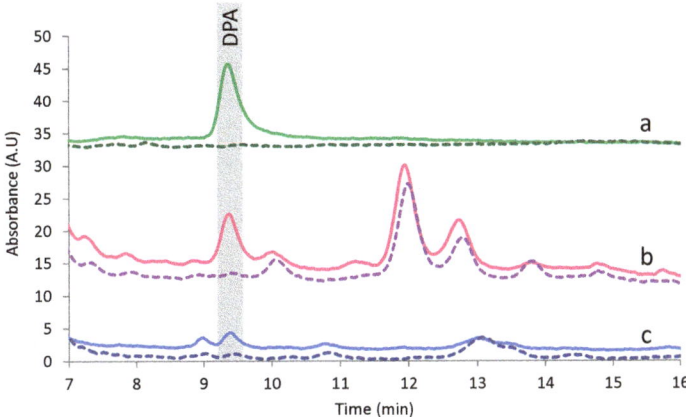

Figure 4. Chromatograms of blanks (dashed lines) and standard solution of 5 ng/mL DPA (solid lines) obtained with different extraction solvents: water (a), 90:10 water: ethanol (b), and 90:10 water: acetonitrile (c). Experimental conditions were the optimized once (see main text for more explanation).

3.3. Analytical Performance of DPA Determination

Relevant analytical parameters such as calibration equations, linear working range, limit of detection (LOD), limit of quantification (LOQ), and precision are shown in Table 3, for both solution and swab-vortex extraction procedures. Satisfactory linearity for the working concentrations was achieved. The LODs and LOQs were calculated experimentally from solutions containing concentrations providing signal/noise of 3 and 10, respectively. Limit of detection and LOQ for the swab-vortex extraction were 0.15 ng/mL and 0.5 ng/mL, respectively. Converted into the equivalent amount of DPA injected onto the system, the LOD and LOQ were 0.3 ng and 1 ng, respectively. These results showed that the sensitivity reached with the proposed procedure is suitable for detecting DPA on shooters' hands and the observed LODs improved the published ones shown in Table 1. The precision was suitable at the working concentration levels tested, with intra- and inter-day relative standard deviations of 9% and 15%, respectively (n = 4). The precision of the retention times was also estimated obtaining RSD values of 1.5% and 2.5% for intra- and inter-day, respectively (n = 3, concentration = 15 ng/mL). Satisfactory results for the study in solution were obtained as depicted in Table 3. To test the extraction efficiency of DPA from samples (including sample collection by cotton swab and extraction from swab to water), the peak area of solution obtained after extraction (2 μL of 10 μg/mL DPA spread on a glass slide followed by the protocol described in Section 2.6) was compared with the peak area obtained for the equivalent concentration in solution directly injected (5 ng/mL of DPA). The extraction efficiency estimated was 37 ± 5%. A recovery study of spiked samples at 10 ng/mL was performed and the value obtained was 108 ± 16%.

Table 3. Analytical data for DPA determination by IT-SPME-CapLC-DAD; a: ordinate, b: slope, s_a and s_b: standard deviation of the ordinate and slope, respectively, R^2: determination coefficient. Limit of detection (LOD) and limit of quantitation (LOQ).

	Linear Range (ng mL^{-1})	$y = a + bx$ (ng mL^{-1})			Precision as % RSD (n = 4, 15 ng mL^{-1})		LOD (ng mL^{-1})	LOQ (ng mL^{-1})
		$a \pm s_a$	$b \pm s_b$	R^2	Intra-Day	Inter-Day		
Solution	0.15–50	−7 ± 57	144 ± 2	0.999	5	10	0.05	0.15
Swab-vortex	0.5–25	−13 ± 24	49.2 ± 1.7	0.994	9	14	0.15	0.5

3.4. Analysis of Samples

Several samples collected from hands of police officers after shooting tests (See Section 2.5) were analyzed by the optimized procedure. Additionally, the same procedure as described for shooting hands (See Section 2.6) was carried out for the hands of each police officer before shooting to obtain blank samples. The samples were analyzed without identification of volunteers. Figure 5 shows the chromatograms for the hands of a shooter (sample 2A) and a non-shooter and the UV-Vis spectra of a standard sample. Diphenylamine was identified in samples by their concordance between retention time (9.4 min) and UV–Vis spectra of DPA from the library. As can be seen in Figure 5, the chromatogram of a blank showed no peak interferences at the retention time of DPA.

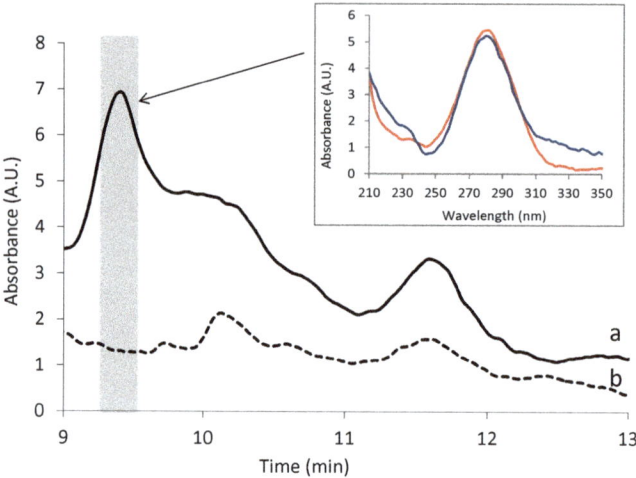

Figure 5. Chromatograms obtained for sample 2A (black solid line) and blank of non-shooter's hand (black dashed line). The inserts correspond to the matching of the spectra of DPA found (blue line) in reference to the standard in the library (red line).

Quantification of the samples was carried out based on the regression equation previously obtained (See Table 3). Table 4 shows the samples screened and the quantification results. With a total of twenty-one swab samples and six tape kit samples, DPA was found and quantified in seventeen swab samples (81% of all swab samples analyzed). In the literature, few studies of DPA are focused on hands and LODs reported are higher to that provided by the proposed method (See Table 1). In this work, the amount of DPA found on hands exceeded LOQ, providing forensic evidence for the presence of DPA. The paired t-test was used to evaluate statistical differences between both hands of a shooter, left and right. The α value obtained at a 95% significant level was higher than 0.05 (p-value = 0.232). From these results, we can conclude that the results from both hands of a shooter were statistically equivalent.

Table 4. Samples screened and quantification of results of DPA on hands determined by the optimized extraction procedure followed by IT-SPME-CapLC-DAD. * Tape lift kit samples quantified by a regression equation with a slope 14 times lower than that obtained for a regression equation by the cotton swab. ** On shooters' hands.

Police Officer	Sample	Hand	Number of Shots	DPA Concentration (ng **)
A	1 A	left	25	4.4
	2 A	right		3.8
B	3 B	left	12	2.7
	4 B	right		1.9
C	5 C	left	25	3.0
	6 C	right		3.8
D	7 D	left	25	2.8
	8 D	right		<LOD
E	9 E	left	25	2.5
	10 E	right		3.2
F	11 F	left	25	16.5
	12 F	right		13.4
G	13 G	left	25	<LOD
	14 G	right		<LOD
H	15 H	left	25	4.9
	16 H	right		<LOD
I	17 I	left	25	8.4
	18 I	right		9.5
J	19 J	left	25	8.0
	20 J	right		4.3
K *	21 K	left	25	<LOD
	22 K	right		<LOD
L	23 L	left and right	25	1.4
M *	24 M	left and right	25	3.6
N *	25 N	left and right	25	6.6
O *	26 O	left and right	25	6.9
P *	27 P	left and right	25	<LOD

3.5. IGSR Particles' Identification

As can be seen in Figure 6, the presence of GSR particles remaining on cotton swabs can be confirmed by naked eye and optical microscopy before chromatographic analysis. Clean fibers of the cotton swab can be seen after sampling a non-shooter's hand (See Figure 6A). However, gunpowder particles with a typical spherical shape and size up to 20 μm [8] were observed between cotton fibers (see red circles) after sampling a shooter's hand (Figure 6B). It is worth mentioning that this non-destructive microscopic analysis allows the subsequent DPA chromatographic analysis too.

Figure 7 shows the same cotton sample (sample 2A) shown in Figure 6 but characterized by SEM/EDX after DPA extraction. This was possible due to the presence of some gunpowder particles remaining on the cotton swabs after the DPA was extracted. Figure 7 shows a typical IGSR particle with a spherical shape and 38 μm size in accordance with References [6,7]. As can be observed in the elemental analysis, the predominant elements were Ba (46%) and Sb (44%), as reported in the literature for IGSRs [4]. Both inorganic and organic compounds were identified on shooters' hands by SEM/EDX and chromatography, respectively. Hence, the presence of GSRs on the hands of shooters was confirmed.

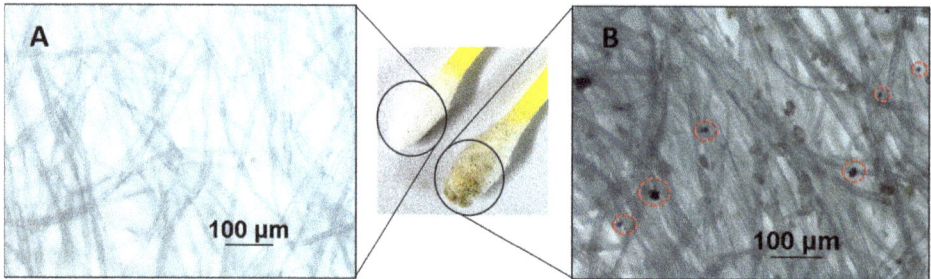

Figure 6. Visual and microscopic (10× magnification) inspection of cotton swab after sampling a non-shooter's hand (**A**) and after sampling a shooter's hand, sample 2A (**B**).

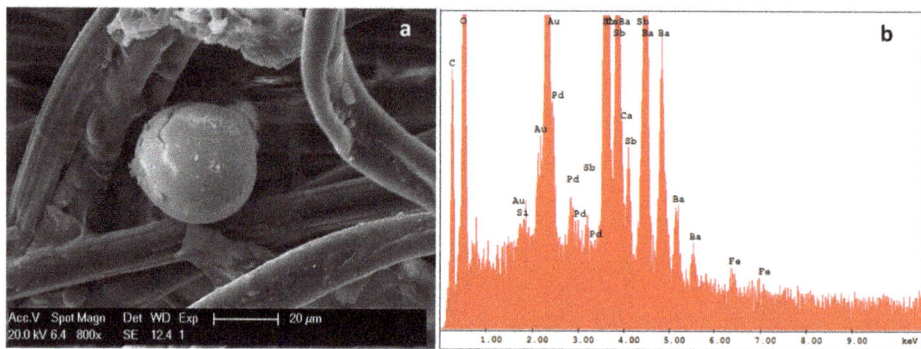

Figure 7. (**a**) Scanning electron microscopy SEM image and (**b**) Energy dispersion X-ray EDX spectra of inorganic gunshot residue found on a swab sample (sample 2A) after shooting.

The other aim of this work was to examine the morphology and elemental composition and distribution of GSR particles collected with the lift tape kits, the typical police collector. Only particles which can be identified as GSR by their composition and morphology were selected for SEM/EDX analysis. Roughly 6–7 particles per sample were studied as can be seen in Table 5. As reported in Reference [8], this number of particles is approximately equivalent to the particles that can be recovered on a shooter's hand at a forensic scene. A portion between 3–40% of the total surface of the sample was explored to find this number of particles, depending on the sample. Figures 8 and 9 show the morphology and elemental data of particles found on adhesive tapes collected after shooting. Most of the particles observed were spherical. Less than 20% of particles found had an irregular shape, probably due to being distorted after shooting. As shown, particles had different surfaces such as smooth, bumpy or covered with craters with or without a metallic shine. More than 60% of the particles found had a smooth surface. Their morphology was an effect of conditions taking place during the firing. Particles can be perforated, capped, broken or stemmed. Results of the SEM/EDX analysis of GRS particles found on the tapes from shooters' hands are displayed in Table 5.

As observed in Table 5, most of the particles had the characteristic elemental composition of GSRs, which was mainly based on Pb, Sb, and Ba; 35 particles contained on average 61% Ba, 30% of Sb, and 9% Pb, and other two particles contained 95% Pb and 97% Sb, probably from bullets, shells or cartridges. Moreover, some particles also contained other elements such as Al, Cu, and Fe at trace levels. About 66% of samples contained traces of Cu, 20% Al, and 3% Fe, while 12% of them contained both Al and Cu. Nevertheless, these minority elements cannot be considered evidence of firing a gun. Even though these particles had similar elemental composition, their size varied over a range from 3 to 30 µm according to the bibliography [4,6,7].

Table 5. Summary of shape, surface, and elemental composition of GRS particles found on tape lift kits from shooters' hands.

Sample	GSR Particle	Shape	Surface	Elemental Composition (%)					
				Major			Minor/Trace		
				Ba	Sb	Pb	Cu	Al	Fe
21K	21K.1	Irregular	Nonmetallic bumpy	62.5	33.2	4.3	X		
	21K.2	Spherical	Nonmetallic smooth	62.4	25.7	11.9	X		
	21K.3	Spherical	Nonmetallic bumpy	65.9	18.8	15.3	X		
	21K.4	Spherical	Nonmetallic bumpy	46.8	40.2	13.0	X		
	21K.5	Spheroidal	Nonmetallic smooth	65.7	14.8	19.5	X		
	21K.6	Spherical	Nonmetallic smooth	87.5	11.1	1.4	X		
	21K.7	Spherical	Nonmetallic bumpy	58.4	28.8	12.8	X		
22K	22K.1	Spherical	Metallic smooth	61.5	33.5	5.0	X		
	22K.2	Irregular	Metallic bumpy	98.8	0.7	0.5			
	22K.3	Spherical	Nonmetallic smooth	48.0	37.3	14.7			
	22K.4	Spherical	Metallic smooth	57.0	37.1	5.9	X		
	22K.5	Spherical	Metallic smooth	61.8	32.7	5.5	X		
	22K.6	Spherical	Metallic smooth	79.0	16.8	4.2	X		
	22K.7	Spherical	Nonmetallic with hollows	50.6	33.4	16.0	X		
24M	24M.1	Spherical	Metallic smooth	63.0	34.8	2.1			
	24M.2	Spherical	Metallic smooth	63.2	30.6	6.3			
	24M.3	Spherical	Nonmetallic bumpy	0.0	96.7	3.3		X	
	24M.4	Spherical	Metallic smooth	60.4	37.2	2.3		X	
	24M.5	Spherical	Nonmetallic smooth	51.1	34.5	14.3	X	X	
	24M.6	Spherical	Metallic smooth	69.2	20.3	10.5	X	X	
	24M.7	Spherical	Metallic smooth	54.8	37.3	7.9	X	X	
25N	25N.1	Spherical	Metallic smooth	68.8	24.7	6.4			X
	25N.2	Irregular	Nonmetallic bumpy	72.0	17.8	10.2	X		
	25N.3	Spheroidal	Metallic bumpy	63.3	30.0	6.7			
	25N.4	Spherical	Metallic smooth	53.0	39.7	7.4	X		
26O	26O.1	Spherical	Metallic smooth	63.6	34.2	2.2			
	26O.2	Spherical	Metallic smooth	61.7	35.2	3.0	X		
	26O.3	Irregular	Nonmetallic bumpy	40.0	33.8	26.2			
	26O.4	Spherical	Metallic smooth	0.3	4.9	94.9	X		
	26O.5	Spherical	Metallic smooth	62.7	31.2	6.3			
	26O.6	Spherical	Metallic smooth	50.6	35.8	13.6			
27P	27P.1	Spherical	Nonmetallic bumpy	53.5	38.3	8.2	X	X	
	27P.2	Spherical	Nonmetallic bumpy	57.4	30.9	11.7	X		
	27P.3	Spherical	Nonmetallic smooth	60.2	36.9	2.9	X		
	27P.4	Spherical	Nonmetallic smooth	57.6	30.9	11.5	X		
	27P.5	Spheroidal	Nonmetallic bumpy	48.9	30.7	20.5	X		
	27P.6	Spherical	Nonmetallic bumpy	61.3	31.7	7.0	X		

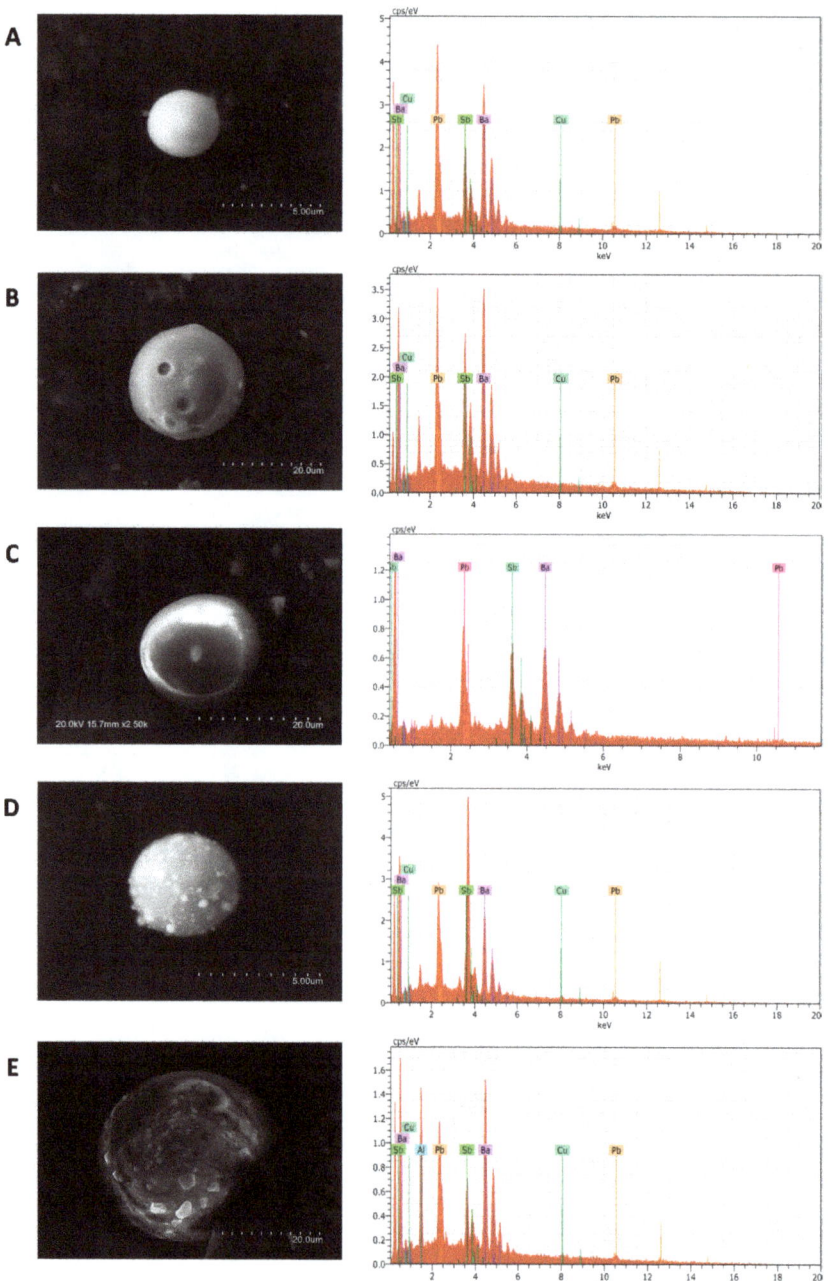

Figure 8. SEM images (left) and EDX spectra (right) of non-spherical particles found on tapes used to collect GSRs after shooting a pistol: sample 22K.2 (**A**), sample 25N.3 (**B**), sample 21K.1 (**C**), sample 25N.2 (**D**).

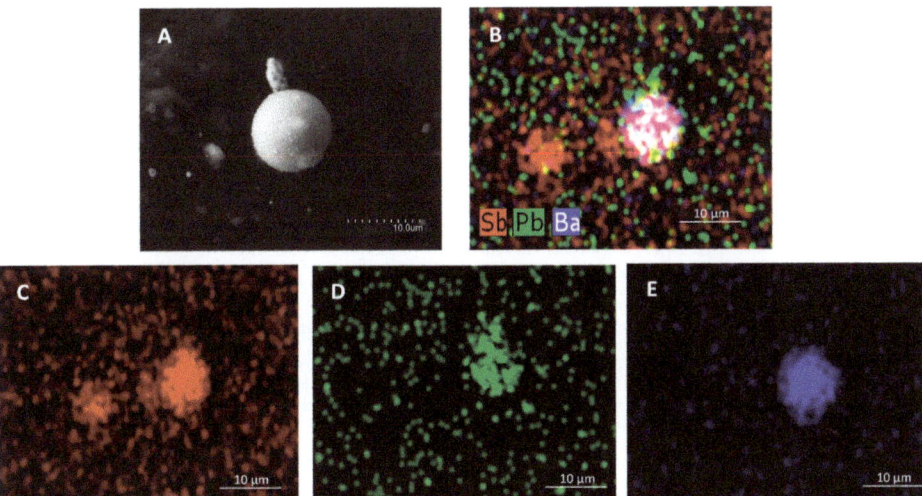

Figure 9. (**A**) SEM image (**B**) overlay X-ray map of singles X-ray of Sb (**C**); Pb (**D**), and Ba (**E**) of particle found on a tape used to collect GSRs on hands after firing a gun.

Spatial distribution of the Sb, Pb, and Ba of GSR particles shown in Table 5 was observed by X-ray mapping using colors to represent the elemental distribution. In this case, Sb appears red, Pb is green, and Ba is blue. Figure 9 shows the X-ray mapping of sample 21K together with its corresponding SEM image. Figure 9B gives the merging of Figure 9C–E. As can be seen, the GSR particle presented the three elements Sb, Ba, and Pb together. Thus, these mapping results were in accordance with the previous elemental composition studied (see Table 5). The results obtained by SEM/EDX can be considered as indicative of IGSR particles on shooters' hands.

4. Conclusions

This work proposes the sampling of gunshot residues on shooters' hands using dry cotton swabs followed by vortex-assisted extraction with water over a short time (20 s). Aqueous samples were directly processed in the miniaturized IT-SPME-CapLC-DAD system for on-line clean-up and preconcentration of the sample and for quantization of the amount of diphenylamine as targeted organic residue. It is worth mentioning that non-toxic solvents and low-cost materials were employed. The efficiencies of the IT-SPME were tested for several compositions and lengths of the extractive phase, as well as sample volume processed in order to improve the sensitivity. The highest analytical responses were obtained for the longest TRB-35 capillaries (90 cm) were more likely due to π–π interactions and 1.8 mL of processed volume. The proposed approach is a rapid, green, and cost-effective option for detecting DPA on the hands of shooters. The sustainability of an analytical method is governed by minimization of toxic solvents, reduction of wastes, and employment of energy-efficient and cost-effective methodologies, but also on maintaining the reliability of the performance parameters, such as sensitivity, precision, and accuracy [25,26]. In two previous papers [27,28] our group demonstrated that IT-SPME-CapLC-DAD achieves the minimization of the sample pre-treatment step, analysis time, and wastes, the reduction of the analysis costs, and thus, improvement of the analytical and environmental performance. Satisfactory LOD (0.3 ng) and precision (RSD intra-day = 9%, RSD inter-day = 14%) were achieved.

In order to test the utility of the method for real cases, several shooters' hands were sampled by dry cotton swabs and processed by IT-SPME-CapLC-DAD. The results showed that DPA was found and quantified in 81% of samples. Additionally, IGSRs inspection of swab samples was carried out by optical microscopy in order to confirm the presence of gunshot residues on shooters' hands, which were analyzed by SEM-EDS after DPA extraction. Furthermore, some shooters' hands were sampled by a tape lift kit, which is the typical police sampler, but DPA extraction was fourteen times lesser than that achieved by the dry cotton swab sampler. Morphology, elemental composition, and distribution of the IGSRs particles were also studied. Then, improved results were obtained by the proposed sampling method as indicated above. If organic compounds are detected in combination with inorganic compounds, higher probative value can be achieved, and false positives/negatives can also be reduced for discriminating shooters' hands. In this work, a sensitive chromatographic method to detect the organic compound DPA can be combined with IGSR analysis by SEM-EDS in order to obtain valuable evidence of GSRs deposited on hands of a suspected shooter. Therefore, the proposed method is helpful to determine whether a person has fired a gun in a forensic investigation.

Author Contributions: All authors designed and performed part of the experiments, analyzed the data, and wrote the paper.

Acknowledgments: The authors are grateful to the EU-FEDER/MINCIU-AEI of Spain (project CTQ2017-90082-P) and Generalitat Valenciana (PROMETEO 2016/109) for the financial support received and to Police Headquarters of Valencian Community (Valencia, Spain) for the sampling.

Conflicts of Interest: The authors declare no conflict of interest. The founding sponsors had no role in the design of the study; in the collection, analyses, or interpretation of data; in the writing of the manuscript, and in the decision to publish the results.

References

1. *Firearms and the Internal Security of the EU: Protecting Citizens and Disrupting Illegal Trafficking*; The European Data Protection Supervisor (EDPS): Brussels, Belgium, 2013.
2. Xu, J.; Murphy, S.L.; Kochanek, K.D.; Bastian, B.A.; Rep, N.V.S. Deaths: Final Data for 2009. *Natl. Vital Stat. Rep.* **2016**, *64*, 1–117.
3. Salles, M.O.; Naozuka, J.; Bertotti, M. A forensic study: Lead determination in gunshot residues. *Microchem. J.* **2012**, *101*, 49–53. [CrossRef]
4. Tarifa, A.; Almirall, J.R. Fast detection and characterization of organic and inorganic gunshot residues on the hands of suspects by CMV-GC-MS and LIBS. *Sci. Justice* **2015**, *55*, 168–175. [CrossRef] [PubMed]
5. Bernal Morales, E.; Revilla Vázquez, A.L. Simultaneous determination of inorganic and organic gunshot residues by capillary electrophoresis. *J. Chromatogr. A* **2004**, *1061*, 225–233. [CrossRef]
6. Dalby, O.; Butler, D.; Birkett, J.W. Analysis of gunshot residue and associated materials—A review. *J. Forensic Sci.* **2010**, *55*, 924–943. [CrossRef] [PubMed]
7. Schwoeble, A.J.; Exline, D.L. *Gunshot Methods in Current Methods in Gunshot Residue Analysis*; CRC Press: London, UK, 2000.
8. Bailey, M.J.; Kirkby, K.J.; Jeynes, C. Trace element profiling of gunshot residues by PIXE and SEM-EDS: A feasibility study. *X-Ray Spectrom.* **2009**, *38*, 190–194. [CrossRef]
9. Brożek-Mucha, Z. Trends in analysis of gunshot residue for forensic purposes. *Anal. Bioanal. Chem.* **2017**, *409*, 5803–5811. [CrossRef] [PubMed]
10. Farokhcheh, A.; Alizadeh, N. Determination of diphenylamine residue in fruit samples using spectrofluorimetry and multivariate analysis. *LWT—Food Sci. Technol.* **2013**, *54*, 6–12. [CrossRef]
11. Grima, M.; Butler, M.; Hanson, R.; Mohameden, A. Firework displays as sources of particles similar to gunshot residue. *Sci. Justice* **2012**, *52*, 49–57. [CrossRef] [PubMed]
12. Oommen, Z.; Pierce, S.M. Lead-free primer residues: A qualitative characterization of Winchester WinCleanTM, Remington/UMC LeadLessTM, Federal BallistiCleanTM, and Speer Lawman CleanFireTM handgun ammunition. *J. Forensic Sci.* **2006**, *51*, 509–519. [CrossRef] [PubMed]
13. Blakey, L.S.; Sharples, G.P.; Chana, K.; Birkett, J.W. Fate and Behavior of Gunshot Residue—A Review. *J. Forensic Sci.* **2017**, *63*, 9–19. [CrossRef] [PubMed]
14. Burleson, G.L.; Gonzalez, B.; Simons, K.; Yu, J.C.C. Forensic analysis of a single particle of partially burnt gunpowder by solid phase micro-extraction-gas chromatography-nitrogen phosphorus detector. *J. Chromatogr. A* **2009**, *1216*, 4679–4683. [CrossRef] [PubMed]

15. Mei, H.; Quan, Y.; Wang, W.; Zhou, H.; Liu, Z.; Shi, H.; Wang, P. Determination of Diphenylamine in Gunshot Residue by HPLC-MS/MS. *J. Forensic Sci. Med.* **2016**, *2*, 18. [CrossRef]
16. Tong, Y.; Wu, Z.; Yang, C.; Yu, J.; Zhang, X.; Yang, S.; Wen, Y. Determination of diphenylamine stabilizer and its nitrated derivatives in smokeless gunpowder using a tandem MS method. *Analyst* **2001**, *126*, 480–484. [CrossRef] [PubMed]
17. Morelato, M.; Beavis, A.; Ogle, A.; Doble, P.; Kirkbride, P.; Roux, C. Screening of gunshot residues using desorption electrospray ionisation-mass spectrometry (DESI-MS). *Forensic Sci. Int.* **2012**, *217*, 101–106. [CrossRef] [PubMed]
18. Laza, D.; Nys, B.; De Kinder, J.; Kirsch-De Mesmaeker, A.; Moucheron, C. Development of a Quantitative LCMS-MS Method for the Analysis of Common Propellant Powder Stabilizers in Gunshot Residue. *J. Forensic Sci.* **2007**, *52*, 842–850. [CrossRef] [PubMed]
19. Moliner-Martinez, Y.; Herráez-Hernández, R.; Verdú-Andrés, J.; Molins-Legua, C.; Campíns-Falcó, P. Recent advances of in-tube solid-phase microextraction. *TrAC–Trends Anal. Chem.* **2015**, *71*, 205–213. [CrossRef]
20. Serra-Mora, P.; Moliner-Martínez, Y.; Molins-Legua, C.; Herráez-Hernández, R.; Verdú-Andrés, J.; Campíns-Falcó, P. Trends in online in-tube solid phase microextraction. In *Comprehensive Analytical Chemistry*; Elsevier: Amsterdam, The Netherlands, 2017; Chapter 14; Volume 76, pp. 427–461.
21. Scherperel, G.; Reid, G.E. Waddell Smith, R. Characterization of smokeless powders using nanoelectrospray ionization mass spectrometry (nESI-MS). *Anal. Bioanal. Chem.* **2009**, *394*, 2019–2028. [CrossRef] [PubMed]
22. Arndt, J.; Bell, S.; Crookshanks, L.; Lovejoy, M.; Oleska, C.; Tulley, T.; Wolfe, D. Preliminary evaluation of the persistence of organic gunshot residue. *Forensic Sci. Int.* **2012**, *222*, 137–145. [CrossRef]
23. Gandy, L.; Najjar, K.; Terry, M.; Bridge, C. A novel protocol for the combined detection of organic, inorganic gunshot residue. *Forensic Chem.* **2018**, *8*, 1–10. [CrossRef]
24. European Chemicals Agency (ECHA). Available online: https://echa.europa.eu/es/information-on-chemicals/registered-substances (accessed on 31 January 2019).
25. Galuszka, A.; Konieczka, P.; Migaszewski, Z.M.; Namiesnik, J. Analytical eco-scale for assessing the greenness of analytical procedures. *TrAC-Trend. Anal. Chem.* **2012**, *37*, 61–72. [CrossRef]
26. Plotka-Wasylka, J. A new tool for the evaluation of the analytical procedures: Green Analytical Procedure Index. *Talanta* **2018**, *181*, 204–209. [CrossRef] [PubMed]
27. Pla-Tolós, J.; Serra-Mora, P.; Hakobyan, L.; Molins-Legua, C.; Moliner-Martinez, Y.; Campins-Falcó, P. A sustainable on-line CapLC method for quantifying antifouling agents like irgarol-1051 and diuron in water samples: Estimation of the carbon footprint. *Sci. Total Environ.* **2016**, *560–570*, 611–618. [CrossRef] [PubMed]
28. Jornet-Martínez, N.; Bocanegra-Rodríguez, S.; González-Fuenzalida, R.A.; Molins-Legua, C.; Campíns-Falcó, P. In Situ Analysis Devices for Estimating the Environmental Footprint in Beverages Industry. In *Processing and Sustainability of Beverages*; Grumezescu, A.M., Holban, A.M., Eds.; Elsevier: Amsterdam, The Netherlands, 2019; Volume 2, Chapter 9; pp. 275–317.

 © 2019 by the authors. Licensee MDPI, Basel, Switzerland. This article is an open access article distributed under the terms and conditions of the Creative Commons Attribution (CC BY) license (http://creativecommons.org/licenses/by/4.0/).

Article

Retention Behaviour of Alkylated and Non-Alkylated Polycyclic Aromatic Hydrocarbons on Different Types of Stationary Phases in Gas Chromatography

Ewa Skoczyńska * and Jacob de Boer

Vrije Universiteit, Department of Environment and Health, De Boelelaan 1108, 1081 HZ Amsterdam, The Netherlands; jacob.de.boer@vu.nl
* Correspondence: e.m.skoczynska@vu.nl

Received: 30 November 2018; Accepted: 24 January 2019; Published: 29 January 2019

Abstract: The gas chromatographic retention behaviour of 16 polycyclic aromatic hydrocarbons (PAHs) and alkylated PAHs on a new ionic liquid stationary phase, 1,12-di(tripropylphosphonium) dodecane bis(trifluoromethanesulfonyl)imide (SLB®-ILPAH) intended for the separation of PAH mixtures, was compared with the elution pattern on more traditional stationary phases: a non-polar phenyl arylene (DB-5ms) and a semi-polar 50% phenyl dimethyl siloxane (SLB PAHms) column. All columns were tested by injections of working solutions containing 20 parental PAHs from molecular weight of 128 to 278 g/mol and 48 alkylated PAHs from molecular weight of 142 to 280 g/mol on a one dimensional gas chromatography-mass spectrometry (GC-MS) system. The SLB PAHms column allowed separation of most isomers. The SLB®-ILPAH column showed a rather different retention pattern compared to the other two columns and, therefore, provided a potential for use in comprehensive two-dimensional GC (GC×GC). The ionic liquid column and the 50% phenyl column showed good thermal stability with a low bleed profile, even lower than that of the phenyl arylene "low bleed" column.

Keywords: ionic liquid stationary phase; gas chromatography; chromatographic selectivity; alkylated polycyclic aromatic hydrocarbons (alkylated PAHs)

1. Introduction

Ubiquitously present in the environment, polycyclic aromatic hydrocarbons (PAHs) originate from natural and anthropogenic incomplete combustion processes. They are present in air, food, water and soil. Nowadays, the PAHs originating from anthropogenic activities are unarguably predominant compared to those originating from natural sources. Humans are exposed to PAHs in almost every aspect of everyday life and, therefore, PAHs are among the most studied chemicals. During the last 50 years, the procedures for the determination of individual PAHs in complex environmental mixtures have been extensively developed and improved. In 1976, 16 specific PAHs were selected for regulation by the United States Environmental Protection Agency (U.S. EPA); the historical perspectives regarding the choice of these 16 EPA PAHs can be found in an article by Keith [1].

In 2002, the toxicities of 33 PAHs were assessed by The European Scientific Committee on Food and 15 PAHs showed clear evidence of mutagenicity/genotoxicity. Fourteen of these 15 PAHs showed clear carcinogenic effects in various types of bioassays and in experimental animals [2]. Seven of these carcinogenic PAHs in the Scientific Committee on Food study are also contained in the EPA's set of 16 PAHs, while the additional seven are: benzo(j)fluoranthene, cyclopenta(cd)pyrene, dibenzo(a,e)-, dibenzo(a,h)-, dibenzo(a,i)-, dibenzo(a,l)pyrene and 5-methylchrysene. In 2006, the Joint FAO/WHO Expert Committee on Food Additives (JECFA) concluded that benzo(c)fluorene is probably also carcinogenic [3]. This shows that the list of the toxic and environmentally relevant PAHs is still growing.

In non-occupational settings, food is the main source of human exposure to PAHs, followed by cigarette smoke, which in some cases may result in PAH exposure on par with the food uptake route [4,5]. Other important exposure routes include traffic related air pollution and all kinds of occupational exposures. Nonetheless, the new possible exposure pathways are still being identified: e.g. synthetic turf materials used on football fields [6].

The analysis of PAHs is generally based on gas chromatography (GC) rather than on liquid chromatography (LC) because GC allows greater selectivity, resolution and sensitivity than LC [7,8]. The GC systems are commonly coupled with flame ionisation detectors (FID) or mass-spectrometric detectors (MS). The GC analysis was conventionally based on non-polar stationary phases operated at relatively high temperatures [8,9]. The 5% phenyl methylpolysiloxane phase (like in the DB-5 column) is still the most often applied one in PAHs analysis and it has also been recommended in a number of US-EPA methods, e.g. US EPA method 610 [10]. Since the 1990s, high phenyl content stationary phases have been more frequently used, e.g. described by the producers as "50% phenyl methylpolysiloxane-like" DB-17MS [8,11], Rxi-PAH [12] or SLB PAHms [13].

Some years ago, a new group of stationary phases, based on non-bonded ionic liquids (IL) was introduced [14,15]. Based on non-molecular solvents with low melting points, these stationary phases consist of organic cations plus inorganic or organic anions [16] and, therefore, the IL columns enable chromatographic separation based on a selectivity different to that provided by conventional stationary phases. Some IL columns can exhibit "dual nature" features; they allow separation of non-polar molecules as non-polar stationary phases do, while at the same time they have a high affinity for polar molecules like polyethylene glycol (wax) and cyanopropyl-siloxane stationary phases. The IL columns are more polar than the wax columns but they have higher thermal stability compared to traditional siloxane phases with a similar selectivity because they are not susceptible to back-biting reactions that result in phase degradation and column bleed [14]. Siloxane-based stationary phases contain active hydroxyl groups at the terminal positions; this makes them sensitive to the oxygen catalyzed cleavage of backbone siloxane. The siloxane chain then breaks to volatile cyclic siloxanes that elute from the column as "bleed" and results in a rising baseline.

So far, the chromatographic properties of the IL columns have only been investigated in a few studies. The IL columns have been used for the separation of different classes of environmental pollutants, like polybrominated diphenyl ethers (PBDEs), polychlorinated biphenyls (PCBs) and other chlorinated compounds [17,18], alkyl phosphates, fatty acids, and petroleum distillates [19]. A new IL column, SLB®-ILPAH, intended for the separation of PAHs mixtures, recently became commercially available. This column has already been tested in terms of the retention behaviour of alkyl-substituted polycyclic aromatic sulphur heterocyclic isomers [13].

In this study, we investigated the retention behaviour of PAHs and alkylated PAHs on the SLB®-ILPAH column and two stationary phases traditionally used for the PAH analysis: a low bleed column with a phenyl arylene polymer that is virtually equivalent to a (5%-phenyl)-methylpolysiloxane (DB-5ms) and a high phenyl content column denoted as 50% phenyl-dimethylpolysiloxane (SLB PAHms). The difference between the arylene column and a 5% phenyl-dimethylsiloxane column is that in the arylene column, the phenyl ring is built in the siloxane chain, whereas in the phenyl-dimethylsiloxane phase, the phenyl rings are positioned as substituents (side chains). Alkylated PAHs were selected because numerous isomers of these compounds are currently targeted in analyses of environmental samples.

Alkylated PAHs are recognised as environmental pollutants although they are still not regularly included in the analysis of priority PACs (e.g. 16 EPA PAHs). They are ubiquitously present in the environment and are often more toxic than the parental PAHs [20,21]. Alkylated PAHs have been found in the toxic fractions in several Effect Directed Analysis (EDA) studies [22–25]. 5-methylchrysene, 1-methylpyrene and 7,12-dimethylbenz(a)anthracene, as confirmed toxic compounds, are being included more and more in standard PAH analyses [26,27]. A list of 34 PAHs (18 parental PAHs and 16 alkylated), has been recommended for toxicological screenings by the US EPA [28]. In addition to the 16 traditional EPA PAHs, the list of 34 PAHs includes perylene, benzo(*e*)pyrene and 16 groups of C-1 to C-4 alkyl derivatives.

The determination of alkylated PAHs in complex environmental samples is problematic because of numerous coeluting isomers [19,29]. It is not possible to separate all isomers of heavier PAHs in a single chromatographic run on one column but two-dimensional GC-MS analysis (GC×GC-MS) could offer a solution. GC×GC can only be fruitful if the two columns used in series are (semi-) orthogonal, or, as chemically different from each other as possible. Therefore, an assessment of new and different stationary phases with different separation mechanisms was needed.

This study investigates the retention behaviour of 20 parental PAHs from molecular weight (MW) 128 to 278 g/mol and 48 alkylated PAHs on three stationary phases. The isomeric sets of alkyl PAHs investigated here are: methyl- and dimethyl-naphthalenes (128-C1, 128-C2), methyl-phenanthrenes and anthracenes (178-C1), methyl-fluoranthene and pyrene (202-C1), methyl- and dimethyl- benz(*a*)anthracenes, benzo(*c*)phenanthrenes and chrysenes (228-C1, 228-C2) and methyl benzo(*a*)pyrenes (252-C1).

2. Materials and Methods

Table 1 shows the characteristics of the three stationary phases that were tested in this study. The columns SLB PAHms and SLB®-ILPAH (both from Supelco, Bellefonte, Pennsylvania, USA) were made available by Sigma Aldrich (Zwijndrecht, The Netherlands) and the DB-5ms was bought from Agilent, The Netherlands. The parental and alkylated-PAHs standard solutions and pure compounds (Table 2) were purchased from Sigma Aldrich (Zwijndrecht, The Netherlands). All solvents used (isooctane and toluene) were obtained in picograde quality from Merck Millipore (Amsterdam, The Netherlands).

Table 1. Stationary phases and their characteristics.

GC Column	Stationary Phase	Dimensions	Max. temp. (Isotherm/ Programmed) °C
VDB-5ms	Phenyl Arylene polymer, virtually equivalent to 5%-phenyl-methylpolysiloxane	30 m × 0.25 mm ID × 0.25 µm	300/320 °C
SLB PAHms (Supelco)	Denoted as 50% phenyl dimethylpolysiloxane	30 m × 0.25 mm ID × 0.25 µm	350/360 °C
SLB®-ILPAH (Supelco)	Non-bonded, 1,12-Di(tripropylphosphonium) dodecane bis(trifluoromethanesulfonyl)imide	20 m × 0.18 mm ID × 0.05 µm	300/300 °C

All standards were gravimetrically prepared in toluene and isooctane. The working solutions were prepared by mixing appropriate volumes from the individual stock solutions. Analyses were performed on an Agilent 6890 gas chromatograph coupled to an Agilent 5975C inert MSD with a Triple-Axis Detector. All injections were performed in the splitless mode (1 µL; splitless time 1.4 min) at 275 °C and with MS operating in total ion current mode. The oven temperature programs were set as follows: DB-5ms and SLB-PAH: isothermal at 90 °C for 10 min, then with 5 °C/min to 300 °C, SLB®-ILPAH: isothermal at 90 °C for 6 min, then with 5 °C/min to 300 °C.

The temperature programs were optimised in order to compare the elution order and peak resolution between the columns.

The SLB®-ILPAH is commercially available in dimensions different from the "standard" dimensions (Table 1) as discussed in the Results and discussion section.

Table 2. PAHs and alkylated PAHs: retention times (RT) and relative retention times (RRT) in minutes. RRTs were calculated relative to pyrene. The coeluting isomers are marked: green (overlap > 90%), blue (90% < overlap > 50%) and orange (overlap < 50%).

Code	DB-5ms	RT	RRT	Code	SLB PAHms	RT	RRT	Code	SLB-ILPAH	RT	RRT
N	Naphthalene	13.20	0.353	N	Naphthalene	16.12	0.379	N	Naphthalene	5.86	0.188
N2	2-Methylnaphthalene	17.65	0.472	N2	2-Methylnaphthalene	20.03	0.471	N2	2-Methylnaphthalene	9.13	0.292
N1	1-Methylnaphthalene	18.19	0.486	N1	1-Methylnaphthalene	20.82	0.489	N1	1-Methylnaphthalene	9.33	0.299
N2,6	2,6-Dimethylnaphthalene	21.25	0.568	N2,6	2,6-Dimethylnaphthalene	23.31	0.548	N2,7	2,7-Dimethylnaphthalene	12.07	0.386
N2,7	2,7-Dimethylnaphthalene	21.31	0.570	N2,7	2,7-Dimethylnaphthalene	23.35	0.549	N2,6	2,6-Dimethylnaphthalene	12.11	0.388
N1,3	1,3-Dimethylnaphthalene	21.66	0.579	N1,3	1,3-Dimethylnaphthalene	24.04	0.565	N1,3	1,3-Dimethylnaphthalene	12.14	0.389
N1,6	1,6-Dimethylnaphthalene	21.78	0.583	N1,6	1,6-Dimethylnaphthalene	24.05	0.565	N1,6	1,6-Dimethylnaphthalene	12.21	0.391
N1,4	1,4-Dimethylnaphthalene	22.22	0.594	N1,4	1,4-Dimethylnaphthalene	24.73	0.581	N1,4	1,4-Dimethylnaphthalene	12.21	0.391
N1,5	1,5-Dimethylnaphthalene	22.31	0.597	N1,5	1,5-Dimethylnaphthalene	24.87	0.584	N1,5	1,5-Dimethylnaphthalene	12.34	0.395
Al	Acenaphthylene	22.53	0.603	N1,2	1,2-Dimethylnaphthalene	25.26	0.593	N1,2	1,2-Dimethylnaphthalene	13.23	0.424
N1,2	1,2-Dimethylnaphthalene	22.65	0.606	Al	Acenaphthylene	25.96	0.610	N1,8	1,8-Dimethylnaphthalene	13.72	0.439
N1,8	1,8-Dimethylnaphthalene	23.25	0.622	N1,8	1,8-Dimethylnaphthalene	26.17	0.615	At	Acenaphthene	13.72	0.439
At	Acenaphthene	23.48	0.628	At	Acenaphthene	26.66	0.626	N1,6,7	1,6,7-Trimethylnaphthalane	15.83	0.507
N1,6,7	1,6,7-Trimethylnaphthalane	25.59	0.684	N1,6,7	1,6,7-Trimethylnaphthalane	27.82	0.654	Al	Acenaphthylene	15.88	0.508
Fl	Fluorene	26.13	0.699	Fl	Fluorene	29.38	0.690	Fl	Fluorene	17.29	0.554
Ph	Phenanthrene	30.75	0.822	Ph	Phenanthrene	34.88	0.819	Ph	Phenanthrene	23.84	0.763
A	Anthracene	30.99	0.829	A	Anthracene	35.10	0.825	A	Anthracene	23.98	0.768
Ph2	2-Methylphenanthrene	33.25	0.889	Ph2	2-Methylphenanthrene	37.22	0.874	45MP	4,5-Methylenephenanthrene	25.49	0.816
An2	2-Methylanthracene	33.47	0.895	An2	2-Methylanthracene	37.38	0.878	Ph2	2-Methylphenanthrene	25.89	0.829
An1	1-Methylanthracene	33.55	0.897	An1	1-Methylanthracene	37.64	0.884	An1	1-Methylanthracene	25.93	0.830
45MP	4,5-Methylenephenanthrene	33.68	0.901	An2	2-Methylanthracene	37.90	0.890	An2	2-Methylanthracene	26.03	0.833
Ph1	1-Methylphenanthrene	33.73	0.902	Ph1	1-Methylanthracene	37.92	0.891	Ph1	1-Methylphenanthrene	26.15	0.837
An9	9-Methylanthracene	34.40	0.920	45MP	4,5-Methylenephenanthrene	38.84	0.912	An9	9-Methylanthracene	26.58	0.851
Ph3,6	3,6-Dimethylphenanthrene	35.36	0.946	An9	9-Methylanthracene	38.87	0.913	Ph3,6	3,6-Dimethylphenanthrene	27.72	0.888
Fa	Fluoranthene	36.39	0.973	Ph3,6	3,6-Dimethylphenanthrene	40.64	0.955	Ph9,10	9,10-Dimethylphenanthrene	28.90	0.925
An2,3	2,3-Dimethylanthracene	36.63	0.980	An2,3	2,3-Dimethylanthracene	41.16	0.967	An2,3	2,3-Dimethylanthracene	29.10	0.932
Py	Pyrene	37.39	1.000	Fa	Fluoranthene	42.36	0.995	Fa	Fluoranthene	30.42	0.974
An9,10	9,10-Dimethylanthracene	37.63	1.006	Py	Pyrene	42.57	1.000	Py	Pyrene	31.23	1.000
Fa2	2-Methylfluoranthene	38.58	1.032	Fa2	2-Methylfluoranthene	43.11	1.013	Fl2	2-Methylfluoranthene	32.29	1.034
Py1	1-Methylpyrene	40.07	1.072	Py1	1-Methylpyrene	45.25	1.063	Bc1	1-Methylbenzo(c)phenanthrene	33.37	1.069
Bc1	1-Methylbenzo(c)phenanthrene	42.36	1.133	Bc1	1-Methylbenzo(c)phenanthrene	47.71	1.121	Py1	1-Methylpyrene	33.43	1.070
Ba	Benz(a)anthracene	43.12	1.153	Bc2	2-Methylbenzo(c)phenanthrene	48.63	1.142	Bc2	2-Methylbenzo(c)phenanthrene	35.98	1.152
T	Triphenylene	43.22	1.156	Ba	Benz(a)anthracene	48.68	1.144	Bc1,12	1,12-Dimethylbenzo(c)phenanthrene	36.46	1.168
C	Chrysene	43.27	1.157	T	Triphenylene	49.00	1.151	Bc4	4-Methylbenzo(c)phenanthrene	36.72	1.176
Bc2	2-Methylbenzo(c)phenanthrene	43.49	1.163	C	Chrysene	49.07	1.153	Bc3	3-Methylbenzo(c)phenanthrene	36.77	1.178
23BA	2,3-Benzanthracene	43.72	1.169	Bc3	3-Methylbenzo(c)phenanthrene	49.42	1.161	Bc5	5-Methylbenzo(c)phenanthrene	36.82	1.179
Bc3	3-Methylbenzo(c)phenanthrene	44.11	1.180	23BA	2,3-Benzanthracene	49.51	1.163	Ba	Benz(a)anthracene	37.29	1.194

Table 2. Cont.

Code	DB-5ms	RT	RRT	Code	SLB PAHms	RT	RRT	Code	SLB-ILPAH	RT	RRT
Bc5	5-Methylbenzo(c)phenanthren	44.39	1.187	Bc5	5-Methylbenzo(c)phenanthren	49.90	1.172	C	Chrysene	37.43	1.199
Bc4	4-Methylbenzo(c)phenanthren	44.45	1.189	Bc4	4-Methylbenzo(c)phenanthren	49.98	1.174	T	Triphenylene	37.56	1.203
Ba2	2-Methylbenz(a)anthracene	44.92	1.201	Ba2	2-Methylbenz(a)anthracene	50.13	1.178	23Ba	2,3-Benzanthracene	37.85	1.212
Ba1	1-Methylbenz(a)anthracene	44.92	1.201	Ba7	7-Methylbenz(a)anthracene	50.39	1.184	Ba1	1-Methylbenz(a)anthracene	37.85	1.212
Ba7	7-Methylbenz(a)anthracene	45.08	1.206	Ba9	9-Methylbenz(a)anthracene	50.47	1.186	C5	5-Methylchrysene	38.41	1.230
Ba9	9-Methylbenz(a)anthracene	45.08	1.206	Ba1	1-Methylbenz(a)anthracene	50.52	1.187	C4	4-Methylchrysene	38.51	1.234
Ba6	6-Methylbenz(a)anthracene	45.16	1.208	Ba6	6-Methylbenz(a)anthracene	50.52	1.187	Ba6	6-Methylbenz(a)anthracene	38.70	1.240
Ba4	4-Methylbenz(a)anthracene	45.16	1.208	Ba4	4-Methylbenz(a)anthracene	50.52	1.187	Ba4	4-Methylbenz(a)anthracene	38.70	1.240
C5	5-Methylchrysene	45.33	1.212	Ba3	3-Methylbenz(a)anthracene	50.96	1.197	Ba2	2-Methylbenz(a)anthracene	38.76	1.242
C6	6-Methylchrysene	45.42	1.215	Ba5	5-Methylbenz(a)anthracene	50.96	1.197	Ba9	9-Methylbenz(a)anthracene	38.85	1.244
Ba3	3-Methylbenz(a)anthracene	45.42	1.215	C6	6-Methylchrysene	51.03	1.199	Ba7	7-Methylbenz(a)anthracene	38.91	1.246
C4	4-Methylchrysene	45.42	1.215	C5	5-Methylchrysene	51.12	1.201	C6	6-Methylchrysene	39.13	1.253
Ba5	5-Methylbenz(a)anthracene	45.42	1.215	C4	4-Methylchrysene	51.25	1.204	Ba3	3-Methylbenz(a)anthracene	39.13	1.253
Bc1,12	1,12-Dimethylbenzo(c)phenanthracene	45.49	1.217	Ba6,8	6,8-Dimethylbenz(a)anthracene	51.26	1.204	Ba5	5-Methylbenz(a)anthracene	39.13	1.253
Ba10	10-Methylbenz(a)anthracene	45.95	1.229	Ba10	10-Methylbenz(a)anthracene	51.73	1.215	Ba10	10-Methylbenz(a)anthracene	39.45	1.264
Ba6,8	6,8-Dimethylbenz(a)anthracene	46.74	1.250	Bc1,12	1,12-Dimethylbenzo(c)phenanthracene	51.91	1.219	Ba6,8	6,8-Dimethylbenz(a)anthracene	39.61	1.269
Ba3,9	3,9-Dimethylbenz(a)anthracene	46.93	1.255	Ba3,9	3,9-Dimethylbenz(a)anthracene	52.17	1.226	Ba7,12	7,12-Dimethylbenz(a)anthracene	39.71	1.272
Ba7,12	7,12-Dimethylbenz(a)anthracene	47.82	1.279	Ba7,12	7,12-Dimethylbenz(a)anthracene	53.98	1.268	Ba3,9	3,9-Dimethylbenz(a)anthracene	40.37	1.293
BbF	Benzo(b)fluoranthene	47.88	1.281	BbF	Benzo(b)fluoranthene	54.02	1.269	Ba8,9,11	8,9,11-Trimethylbenz(a)anthracene	41.40	1.326
BkF	Benzo(k)fluoranthene	47.94	1.282	BkF	Benzo(k)fluoranthene	54.12	1.271	BbF	Benzo(b)fluoranthene	42.86	1.373
BeP	Benzo(e)pyrene	48.88	1.307	Ba8,9,11	8,9,11-Trimethylbenz(a)anthracene	54.30	1.276	BkF	Benzo(k)fluoranthene	43.02	1.378
Ba8,9,11	8,9,11-Trimethylbenz(a)anthracene	49.03	1.311	BeP	Benzo(e)pyrene	55.61	1.306	BeP	Benzo(e)pyrene	44.04	1.411
BaP	Benzo(a)pyrene	49.08	1.313	BaP	Benzo(a)pyrene	55.86	1.312	BaP	Benzo(a)pyrene	44.08	1.412
BaP9	9-Methylbenzo(a)pyrene	50.71	1.356	BaP9	9-Methylbenzo(a)pyrene	57.19	1.343	BaP10	10-Methylbenzo(a)pyrene	45.24	1.449
BaP8	8-Methylbenzo(a)pyrene	50.84	1.360	BaP8	8-Methylbenzo(a)pyrene	57.39	1.348	BaP9	9-Methylbenzo(a)pyrene	45.49	1.457
BaP7	7-Methylbenzo(a)pyrene	51.07	1.366	BaP7	7-Methylbenzo(a)pyrene	57.71	1.356	BaP8	8-Methylbenzo(a)pyrene	45.49	1.457
BaP10	10-Methylbenzo(a)pyrene	51.12	1.367	BaP10	10-Methylbenzo(a)pyrene	57.94	1.361	BaP7	7-Methylbenzo(a)pyrene	45.49	1.457
BaP7,10	7,10-Dimethylbenzo(a)pyrene	52.93	1.416	BaP7,10	7,10-Dimethylbenzo(a)pyrene	59.77	1.404	BaP7,10	7,10-Dimethylbenzo(a)pyrene	46.27	1.482
	Indeno(1,2,3-c,d)pyrene	53.26	1.424		Indeno(1,2,3-c,d)pyrene	60.86	1.430		Dibenz(a,h)anthracene	48.68	1.559
	Dibenz(a,h)anthracene	53.44	1.429		Dibenz(a,h)anthracene	60.94	1.432		Indeno(1,2,3-c,d)pyrene	49.02	1.570
	Benzo(g,h,i)perylene	54.26	1.451		Benzo(g,h,i)perylene	62.89	1.477		Benzo(g,h,i)perylene	50.01	1.602

3. Results and discussion

The retention times and the relative retention times to pyrene of all parental PAHs and alkylated PAHs injected on the three columns are presented in Table 2. This table also shows the coelutions of the isomers having similar mass spectra (see coloured cells). The coelutions of PAHs that are not isomers (e.g. the coelution of 7,12-dimethylbenz(*a*)anthracene and benzo(*b*)fluoranthene or 2-methylbenzo(*c*)phenanthrene and benz(*a*)anthracene on the SLB PAHms column) were not marked here because these compounds have different mass spectra and can be separated by the MS detector. However, the interferences of the fragment ions of the overlapping compounds with different base peak ions must be taken into account for accurate quantitation.

The elution order of the PAHs and the alkyl-PAHs on the phenyl arylene and the 50% phenyl-polysiloxane stationary phases is rather similar. However, the elution order on the SLB-ILPAH is different; these differences will be discussed below. The advantages and shortcomings of the three studied columns are briefly summarized in Table 3.

Table 3. Chromatographic characteristics of the three columns: DB-5ms, SLB PAHms and SLB-ILPAH.

GC Columns	Phenyl Arylene	50% Phenyl Polysiloxane	SLB-ILPAH
Overlap > 90%	12 peaks	11 peaks	19 peaks
90% > overlap > 50%	7 peaks	2 peaks	3 peaks
Overlap < 50%	4 peaks	4 peaks	1 peak
Peak shape	Good	Good	Good
Analysis time	Long	Long	Shorter than on the other two columns
Bleeding	Substantial bleeding above 260 °C	No bleeding till 300 °C	No bleeding till 300 °C

The least polar column, phenyl arylene, shows an overlap of 19 isomers at more than 50% of the peak height and of 4 isomers at less than 50% of the peak height. Chrysene, one of the 16 EPA PAHs, coelutes with triphenylene but the rest of the 16 EPA PAHs are totally resolved. This column showed the best separation of dimethylnaphthalenes (Figure 1); 1,3- and 1,6-dimethylnaphthalenes were separated on this column only. Figure 1 shows that the dimethylnaphthalenes formed a co-eluting peaks' cluster on the ionic liquid column while on the siloxane-based columns they were much better separated. Figure 1 also shows that compared to the phenyl arylene column, the 50% phenyl-polysiloxane column shows a substantially better separation of the injected isomers. Table 2 shows that on this column only 13 isomers overlapped at more than 50% of the peak height and four isomers overlapped at less than 50% of the peak height.

Figure 2A shows that chrysene and triphenylene were partly separated on the 50% phenyl-polysiloxane column while they coeluted at the phenyl arylene column. The separation of these isomers is comparable to the separation achieved on the Rxi-PAH column (50% phenyl methylpolysiloxane-like phase) used for the PAHs analysis by Nalin et al. [12]. The study of Poster et al. [8] showed that chrysene and triphenylene coelute on the comparable DB-17MS stationary phase (50% phenyl methyl-polysiloxane-like phase), are partly resolved on the non-polar DB-XLB column (proprietary phase) and totally resolved on the LC-50 column (dimethyl/50% liquid crystalline phase). Figure 2A shows that chrysene and triphenylene are totally resolved on the IL column.

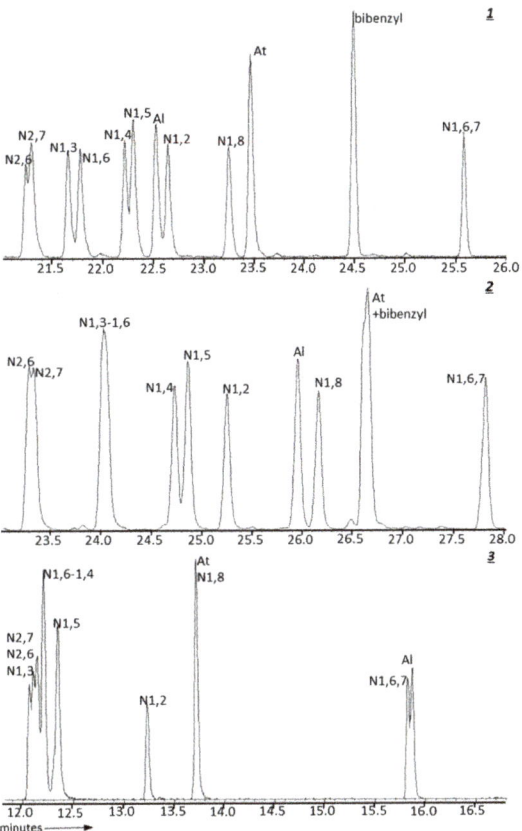

Figure 1. Elution order of dimethylnaphthalenes, trimethylnaphthalene, acenaphthylene (Al), acenaphthene (At) on phenyl arylene (**1**), 50% phenyl-polysiloxane (**2**) and SLB-ILPAH (**3**) stationary phases. For abbreviations see Table 2.

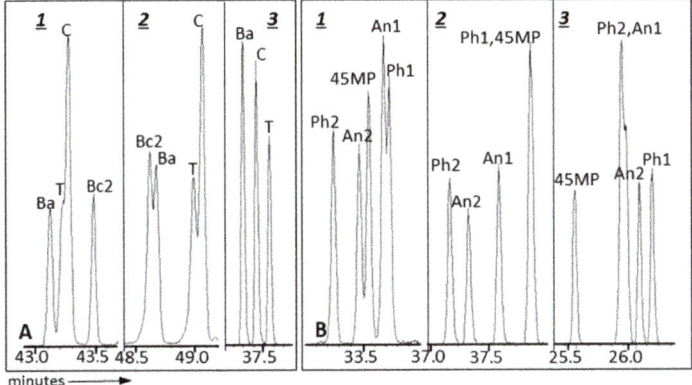

Figure 2. Elution order of 228-PAHs (**A**) and ethyl- anthracenes and phenanthrenes and 4,5-methylenephenanthrene (**B**) on phenyl arylene (**1**), 50% phenyl-polysiloxane (**2**) and SLB-ILPAH (**3**) stationary phases. For abbreviations see Table 2.

In Figure 2B we see that all isomers of methylated phenanthrenes and anthracenes were separated on the 50% phenyl-polysiloxane column, while some of these isomers coeluted on the phenyl arylene and on the SLB-ILPAH column. The IL column also demonstrated the different mechanism of retention; 4,5-methylenephenanthrene eluted before the methylated phenanthrenes and anthracenes (178-C1).

Figure 3 shows that the best separation of 17 methylated benz(*a*)anthracenes, benzo(*c*)phenanthrenes and chrysenes isomers (228-C1) was achieved on the 50% phenyl-polysiloxane column; only seven isomers coeluted at more than 90% of the peak height while the remaining 10 isomers were at least partly resolved (Table 2). This separation was better than the separation achieved on the phenyl arylene column, where 11 of these isomers coeluted, as well as on the SLB-ILPAH column, where 10 of these isomers coeluted. The number of the observed coelutions might be reduced by increasing the lengths of the tested columns, reducing the internal diameters and/or by improving the applied temperature programs with stable temperature periods around the elution times of isomeric clusters.

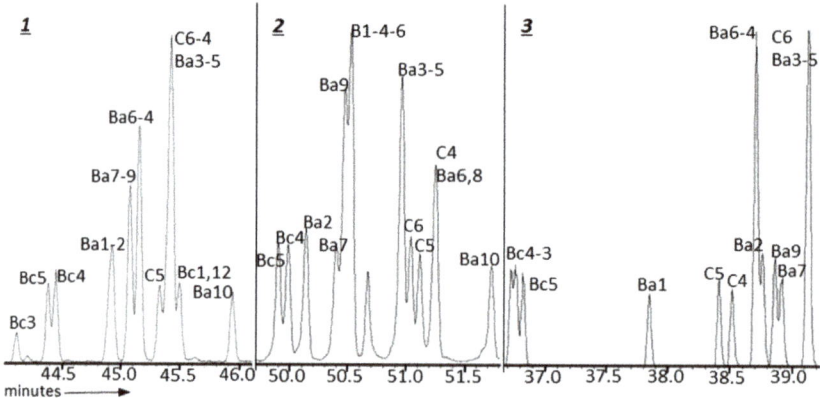

Figure 3. Elution order of methyl-benz(*a*)anthracenes, chrysenes and benzo(*c*)phenanthrenes and dimethylbenzo(*c*)phenanthrenes on phenyl arylene (**1**), 50% phenyl-polysiloxane (**2**) and SLB-ILPAH (**3**) stationary phases. For abbreviations see Table 2.

The commercially available SLB-ILPAH column was 2/3 the length of the two other columns. The internal diameter was 3/4 of that of the other two and the film thickness 1/5 of that of the two other columns, which made the separation substantially faster. However, it is not possible to compare the dimensions of the IL column to the siloxane-based columns directly because of the different nature of an IL coating resulting in the different type of interactions between the analytes and the stationary phase. This IL phase shows stronger retention for heavier PAHs (Table 2); the relative retention times of the heavier PAHs on this column are higher than on the phenyl arylene and the 50% phenyl-polysiloxane columns. The SLB-ILPAH phase also showed some interesting elution shifts: 1,12-dimethylbenzo(*c*)phenanthrene (228-C2) eluted before benz(*a*)anthracene and other PAHs with MWs of 228 g/mol and 1-methylbenzo(*c*)phenanthrene (228-C1) eluted before 1-methylpyrene (202-C1). Also, the elution order of four PAHs from the 16 EPA PAHs-group on this IL column is different compared to the elution on the two siloxane-based columns: acenaphthylene elutes before acenaphthene and dibenz(*a,h*)anthracene elutes before indeno(1,2,3-*cd*)perylene on the SLB-ILPAH column. However, the overall separation of the isomers on the SLB-ILPAH phase is not as good as on the other two phases: 22 isomers overlap at more than 50% of the peak height. A huge advantage of this column is the total separation of chrysene from triphenylene (Figure 2A). Yet, Figure 4 shows that the highly carcinogenic benzo(*a*)pyrene, another PAH belonging to the group of the 16 EPA PAHs, coeluted with benzo(*e*)pyrene. Both isomers are separated on the phenyl-siloxane column, while benzo(*a*)pyrene coelutes with 8,9,11-trimethylbenz(*a*)anthracene. Priority toxicant

5-methylchrysene was totally separated on this column while on the other two columns it could not be totally resolved from other isomers (Figure 3). The SLB-ILPAH column also managed to separate 1-methylbenz(*a*)anthracene and 4-methylchrysene; these isomers (partially) coelute on the other two columns. It is plausible that increasing the length of this column to 30 m may somewhat improve the observed coelutions, but it is unlikely the pattern would improve so much that it would equal that of the other two columns.

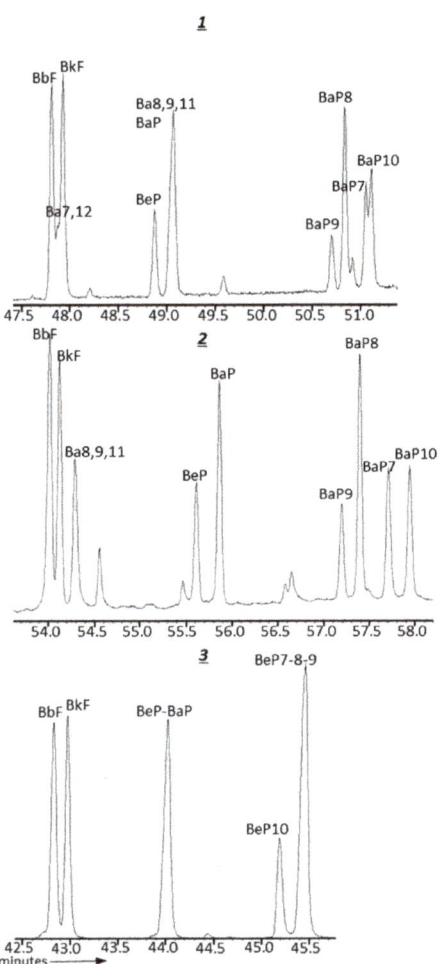

Figure 4. Elution order of benzo(*b*)fluoranthene, benzo(*k*)fluoranthene, benzo(*e*)pyrene, benzo(*a*)pyrene and methylbenzo(*a*)pyrenes on phenyl arylene (**1**), 50% phenyl-polysiloxane (**2**) and SLB-ILPAH (**3**) stationary phases. For abbreviations see Table 2.

Overlap of 3-methylbenz(*a*)anthracene with 5-methylbenz(*a*)anthracene and 4-methylbenz(*a*)anthracene with 6-methylbenz(*a*)anthracene was observed on all three columns (Figure 3). It is worth noting that these isomers could not be separated by GC×GC-MS with different column combinations either [29]. The DB-5 (60 m)×LC-50 (1.2 m) column combination tested by Skoczynska et al. [29] in the analysis of the 228-C1 methylated PAHs was able to separate in the second dimension 7-methylbenz(*a*)anthracene from 9-methylbenz(*a*)anthracene isomers, two compounds that coelute on the DB-5ms and the 50% phenyl-polysiloxane. Significant differences in selectivity between the LC-50 and the Rxi-PAH (50% phenyl comparable to the SLB PAHms phase) were shown in the study of Nalin et al. [12]. The elution pattern of methylchrysenes (228-C1) and methylbenzo(*a*)pyrenes (252-C1) obtained on Rxi-PAH by Nalin et al. is similar to the pattern obtained on the 50% phenyl-polysiloxane in this study (even though Nalin et al. analysed more isomers). Coupling of LC-50 in the second dimension with 50% phenyl-polysiloxane in the first dimension could, therefore, result in orthogonal separation of the coeluting isomers (e.g. 7-methylbenz(*a*)anthracene from 9-methylbenz(*a*)anthracene). The SLB-ILPAH shows the strongest deviation in the retention pattern due to a different type of interactions between the analytes and the stationary phase than in the other two columns studied. Therefore, using this column together with 50% phenyl-polysiloxane may result in orthogonal separation of different PAHs isomers in one GC×GC run. Because of the "dual nature" of the IL columns, the coupling of a "standard" 50%-phenyl polysiloxane column with an IL column in a GC×GC analysis will almost certainly result in an improved separation of the PAHs isomers; a follow-up study may include the evaluation of ionic liquid stationary phases with different polarity coupled to a 50% phenyl-polysiloxane column.

Very little tailing was observed and the peak shapes obtained on all three columns were satisfactory (Figures 1–4). The variation in response obtained on the three columns was relatively small.

Figure 5 shows the column bleed of the three phases: the bleeding of the 50% phenyl-polysiloxane and the SLB-ILPAH phases were comparable and several times lower than the bleeding of the phenyl arylene "low bleed" stationary phase.

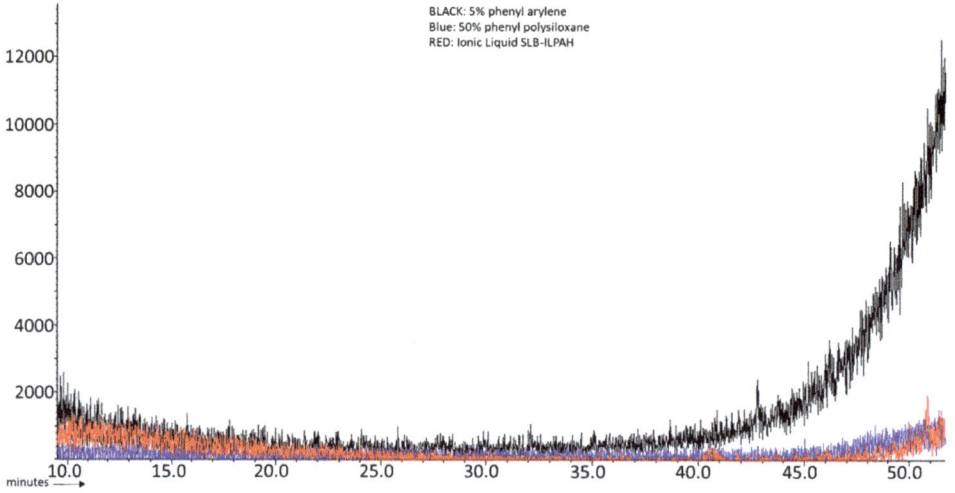

Figure 5. Bleeding of three columns (T max = 300): phenyl arylene (black), 50% phenyl-polysiloxane (blue) and SLB-ILPAH (red) stationary phases.

4. Conclusion

None of the three columns tested offers a complete separation of the injected PAH and methyl-PAH isomers. On the SLB-ILPAH column 22, isomers overlapped at more than 50% of the peak height. The phenyl arylene column showed an overlap of 19 isomers and the 50% phenyl-polysiloxane phase of 13 isomers. Also, none of the columns was able to totally resolve all 16 EPA PAHs. The 50% phenyl-polysiloxane column showed the best overall resolving power and is, therefore, currently considered the best option for the PAH and methyl-PAH analysis.

However, the SLB-ILPAH column is interesting because of a strongly deviating elution pattern, which is due to the different type of interactions between the analytes and the stationary phase.

That makes the ionic liquid column interesting for specific separations that cannot be obtained on one of the other two columns or possibly on other traditional phases. A huge advantage of the ionic liquid column is, for example, the total separation of chrysene from triphenylene. An additional advantage is that using this ionic liquid phase, together with e.g. the 50% phenyl-polysiloxane phase, may result in a (semi-)orthogonal separation of PAHs and methyl PAHs in one GC×GC run.

The ionic liquid SLB-ILPAH column and the high phenyl content 50% phenyl-polysiloxane column both show better thermal stability with less bleeding compared to that of the phenyl arylene "low bleed" column. This low bleeding is an asset for GC×GC because often, more polar columns are used, which show higher bleeding.

Author Contributions: Conceptualization, J.B. and E.S.; methodology, J.B. and E.S.; validation, J.B. and E.S.; formal analysis, E.S.; investigation, E.S.; resources, J.B.; data curation, E.S.; writing—original draft preparation, E.S.; writing—review and editing, J.B. and E.S.; visualization, E.S.; supervision, J.B.

Funding: The research received no external funding.

Acknowledgments: The authors thank Sigma-Aldrich for making the ionic liquid column available.

Conflicts of Interest: The authors declare no conflict of interests.

References

1. Keith, L.H. The Source of US EPA's Sixteen PAH Priority Pollutants. *Polycycl. Aromat. Compd.* **2015**, *35*, 147–160. [CrossRef]
2. Available online: https://ec.europa.eu/food/sites/food/files/safety/docs/sci-com_scf_out153_en.pdf (accessed on 28 January 2019).
3. Available online: http://apps.who.int/iris/bitstream/handle/10665/43258/WHO_TRS_930_eng.pdf?sequence=1 (accessed on 28 January 2019).
4. Ding, Y.S.; Trommel, J.S.; Yan, X.Z.J.; Ashley, D.; Watson, C.H. Determination of 14 polycyclic aromatic hydrocarbons in mainstream smoke from domestic cigarettes. *Environ. Sci. Technol.* **2005**, *39*, 471–478. [CrossRef] [PubMed]
5. Srogi, K. Monitoring of environmental exposure to polycyclic aromatic hydrocarbons: A review. *Environ. Chem. Lett.* **2007**, *5*, 169–195. [CrossRef] [PubMed]
6. Celeiro, M.; Dagnac, T.; Llompart, M. Determination of priority and other hazardous substances in football fields of synthetic turf by gas chromatography-mass spectrometry: A health and environmental concern. *Chemosphere* **2018**, *195*, 201–211. [CrossRef] [PubMed]
7. De Boer, J.; Law, R.J. Developments in the use of chromatographic techniques in marine laboratories for the determination of halogenated contaminants and polycyclic aromatic hydrocarbons. *J. Chromatogr. A* **2003**, *1000*, 223–251. [CrossRef]
8. Poster, D.L.; Schantz, M.M.; Sander, L.C.; Wise, S.A. Analysis of polycyclic aromatic hydrocarbons (PAHs) in environmental samples: A critical review of gas chromatographic (GC) methods. *Anal. Bioanal. Chem.* **2006**, *386*, 859–881. [CrossRef] [PubMed]
9. Wise, S.A.; Sander, L.C.; Schantz, M.M. Analytical Methods for Determination of Polycyclic Aromatic Hydrocarbons (PAHs)—A Historical Perspective on the 16 US EPA Priority Pollutant PAHs. *Polycycl. Aromat. Compd.* **2015**, *35*, 187–247. [CrossRef]
10. Available online: https://www.epa.gov/sites/production/files/2015-10/documents/method_610_1984.pdf (accessed on 28 January 2019).
11. Gomez-Ruiz, J.A.; Wenzl, T. Evaluation of gas chromatography columns for the analysis of the 15+1 EU-priority polycyclic aromatic hydrocarbons (PAHs). *Anal. Bioanal. Chem.* **2009**, *393*, 1697–1707. [CrossRef]
12. Nalin, F.; Sander, L.C.; Wilson, W.B.; Wise, S.A. Gas chromatographic retention behavior of polycyclic aromatic hydrocarbons (PAHs) and alkyl-substituted PAHs on two stationary phases of different selectivity. *Anal. Bioanal. Chem.* **2018**, *410*, 1123–1137. [CrossRef]
13. Wilson, W.B.; Sander, L.C.; Ona-Ruales, J.O.; Moessner, S.G.; Sidisky, L.M.; Lee, M.L.; Wise, S.A. Retention behavior of alkyl-substituted polycyclic aromatic sulfur heterocycle isomers in gas chromatography on stationary phases of different selectivity. *J. Chromatogr. A* **2017**, *1484*, 73–84. [CrossRef]

14. Anderson, J.L.; Armstrong, D.W. Immobilized ionic liquids as high-selectivity/high-temperature/high-stability gas chromatography stationary phases. *Anal. Chem.* **2005**, *77*, 6453–6462. [CrossRef]
15. Poole, C.F.; Poole, S.K. Ionic liquid stationary phases for gas chromatography. *J. Sep. Sci.* **2011**, *34*, 888–900. [CrossRef] [PubMed]
16. Berthod, A.; Ruiz-Angel, M.J.; Huguet, S. Nonmolecular solvents in separation methods: Dual nature of room temperature ionic liquids. *Anal. Chem.* **2005**, *77*, 4071–4080. [CrossRef] [PubMed]
17. De Boer, J.; Blok, D.; Ballesteros-Gomez, A. Assessment of ionic liquid stationary phases for the determination of polychlorinated biphenyls, organochlorine pesticides and polybrominated diphenyl ethers. *J. Chromatogr. A* **2014**, *1348*, 158–163. [CrossRef]
18. Ros, M.; Escobar-Arnanz, J.; Sanz, M.L.; Ramos, L. Evaluation of ionic liquid gas chromatography stationary phases for the separation of polychlorinated biphenyls. *J. Chromatogr. A* **2018**, *1559*, 156–163. [CrossRef]
19. Antle, P.M.; Zeigler, C.D.; Wilton, N.M.; Robbat, A. A more accurate analysis of alkylated PAH and PASH and its implications in environmental forensics. *Int. J. Environ. Anal. Chem.* **2014**, *94*, 332–347. [CrossRef]
20. Stout, S.A.; Emsbo-Mattingly, S.D.; Douglas, G.S.; Uhler, A.D.; McCarthy, K.J. Beyond 16 Priority Pollutant PAHs: A Review of PACs used in Environmental Forensic Chemistry. *Polycycl. Aromat. Compd.* **2015**, *35*, 285–315. [CrossRef]
21. Lam, M.M.; Bulow, R.; Engwall, M.; Giesy, J.P.; Larsson, M. Methylated PACs Are More Potent Than Their Parent Compounds: A Study of Aryl Hydrocarbon Receptor-Mediated Activity, Degradability, and Mixture Interactions in the H4IIE-luc Assay. *Environ. Toxicol. Chem.* **2018**, *37*, 1409–1419. [CrossRef]
22. Brack, W.; Schirmer, K.; Erdinger, L.; Hollert, H. Effect-directed analysis of mutagens and ethoxyresorufin-O-deethylase inducers in aquatic sediments. *Environ. Toxicol. Chem.* **2005**, *24*, 2445–2458. [CrossRef] [PubMed]
23. Kaisarevic, S.; Luebcke-von Varel, U.; Orcic, D.; Streck, G.; Schulze, T.; Pogrmic, K.; Teodorovic, I.; Brack, W.; Kovacevic, R. Effect-directed analysis of contaminated sediment from the wastewater canal in Pancevo industrial area, Serbia. *Chemosphere* **2009**, *77*, 907–913. [CrossRef]
24. Meyer, W.; Seiler, T.-B.; Christ, A.; Redelstein, R.; Puettmann, W.; Hollert, H.; Achten, C. Mutagenicity, dioxin-like activity and bioaccumulation of alkylated picene and chrysene derivatives in a German lignite. *Sci. Total Environ.* **2014**, *497*, 634–641. [CrossRef]
25. Xiao, H.; Krauss, M.; Floehr, T.; Yan, Y.; Bahlmann, A.; Eichbaum, K.; Brinkmann, M.; Zhang, X.; Yuan, X.; Brack, W. Effect-Directed Analysis of Aryl Hydrocarbon Receptor Agonists in Sediments from the Three Gorges Reservoir, China. *Environ. Sci. Technol.* **2016**, *50*, 11319–11328. [CrossRef] [PubMed]
26. Arp, H.P.H.; Azzolina, N.A.; Cornelissen, G.; Hawthorne, S.B. Predicting Pore Water EPA-34 PAH Concentrations and Toxicity in Pyrogenic-Impacted Sediments Using Pyrene Content. *Environ. Sci. Technol.* **2011**, *45*, 5139–5146. [CrossRef] [PubMed]
27. Richter-Brockmann, S.; Achten, C. Analysis and toxicity of 59 PAH in petrogenic and pyrogenic environmental samples including dibenzopyrenes, 7H-benzo c fluorene, 5-methylchrysene and 1-methylpyrene. *Chemosphere* **2018**, *200*, 495–503. [CrossRef]
28. Available online: https://clu-in.org/conf/tio/porewater1/resources/EPA-ESB-Procedures-PAH-mixtures.pdf (accessed on 28 January 2019).
29. Skoczynska, E.; Leonards, P.; de Boer, J. Identification and quantification of methylated PAHs in sediment by two-dimensional gas chromatography/mass spectrometry. *Anal. Methods* **2013**, *5*, 213–218. [CrossRef]

© 2019 by the authors. Licensee MDPI, Basel, Switzerland. This article is an open access article distributed under the terms and conditions of the Creative Commons Attribution (CC BY) license (http://creativecommons.org/licenses/by/4.0/).

Article

Multiple-stage Precursor Ion Separation and High Resolution Mass Spectrometry toward Structural Characterization of 2,3-Diacyltrehalose Family from *Mycobacterium tuberculosis*

Cheryl Frankfater [1,†], Robert B. Abramovitch [2,†], Georgiana E. Purdy [3,†], John Turk [1], Laurent Legentil [4], Loïc Lemiègre [4] and Fong-Fu Hsu [1,*]

1. Department of Medicine, Washington University School of Medicine, St. Louis, MO 63110, USA; c.frankf@wustl.edu (C.F.); jturk@wustl.edu (J.T.)
2. Department of Microbiology and Molecular Genetics, Michigan State University, East Lansing, MI 48824, USA; abramov5@msu.edu
3. Department of Molecular Microbiology and Immunology, Oregon Health & Science University, Portland, OR 97239, USA; purdyg@ohsu.edu
4. Univ Rennes, Ecole Nationale Supérieure de Chimie de Rennes, CNRS, ISCR–UMR 6226, F-35000 Rennes, France; laurent.legentil@ensc-rennes.fr (L.L.); loic.lemiegre@ensc-rennes.fr (L.L.)
* Correspondence: fhsu@im.wustl.edu; Tel.: +1-314-362-0056
† These authors contributed equally as co-first author.

Received: 1 December 2018; Accepted: 7 January 2019; Published: 15 January 2019

Abstract: Mass spectrometry (MS)-based precursor ion isolation, collision-induced dissociation (CID) fragmentation, and detection using linear ion-trap multiple-stage mass spectrometry (LIT MS^n) in combination with high resolution mass spectrometry (HRMS) provides a unique tool for structural characterization of complex mixture without chromatographic separation. This approach permits not only separation of various lipid families and their subfamilies, but also stereoisomers, thereby, revealing the structural details. In this report, we describe the LIT MS^n approach to unveil the structures of a 2,3-diacyl trehalose (DAT) family isolated from the cell envelope of *Mycobacterium tuberculosis*, in which more than 30 molecular species, and each species consisting of up to six isomeric structures were found. LIT MS^n performed on both [M + Na]$^+$ and [M + HCO$_2$]$^-$ ions of DAT yield complimentary structural information for near complete characterization of the molecules, including the location of the fatty acyl substituents on the trehalose backbone. This latter information is based on the findings of the differential losses of the two fatty acyl chains in the MS^2 and MS^3 spectra; while the product ion spectra from higher stage LIT MS^n permit confirmation of the structural assignment.

Keywords: tandem mass spectrometry; linear ion trap; glycolipid; diacyltrehalose; *Mycobacterium tuberculosis*

1. Introduction

Tandem mass spectrometry is a powerful tool for structural analysis of unknown molecules. Tandem mass spectrometry consists of several sequential events including formation and mass selection of the precursor ions, collision induced dissociation (CID) with inert target gas for fragment ion formation, followed by mass analysis and detection of the product-ions. For tandem mass spectrometry with quadrupole (e.g., triple quadrupole, TSQ), sector, and hybrid Q-TOF instruments, these processes occur sequentially in the separate regions of the instruments and the MS/MS process is tandem-in-space. For quadrupole ion-trap (QIT) and linear ion trap (LIT) instruments, the precursor ion selection, CID, and product-ion analysis and detection events are all executed in the ion trap in a timing sequence manner, and the MS/MS process is tandem-in-time [1]. Linear ion trap multiple stage tandem mass spectrometry (LIT MS^n) permits the repeat of the precursor ion selection-CID-product-ion analysis process, and up to 10 cycles can be performed using modern commercial LIT mass spectrometers such as Thermo LTQ Orbitrap. This instrument is also featured with an extreme high resolving power and up to a million resolution (at m/z 200) can be reached [2].

Due to its MS^n capability with high resolution, LIT/Orbitrap with MS^n approach has been widely used in the structural characterization of a wide range of biomolecules [3–7]. LIT MS^n with high resolution is also extremely useful for identification of complex lipid structures, in particular, microbial lipid, permitting revelation of numerous isomeric molecules in lipid extracts. For example, we demonstrated that sulfolipid II in *Mycobacterium tuberculosis* (*M. tuberculosis*) H37Rv cells is the predominated lipid family, consisting of hundreds of molecular species rather than the sulfolipid I family as previously reported [8,9]. LIT MS^n with high resolution mass spectrometry has also been successfully applied to delineate the structures of phosphatidylinositol mannosides (PIMs) [10,11], and phthiocerol dimycocerosates (PDIMs) [12] in the cell envelope of *M. tuberculosis*. The former lipid family is known to have played important roles in *M. tuberculosis* adhesins that mediate attachment to nonphagocytic cells [13], while the latter is recently found to play role in drug resistance to *M. tuberculosis* [14].

In addition to the above complex lipid families, other glycolipids found in the mycobacterial cell wall include acylated trehaloses [15–18]. These trehalose-containing glycolipids consist of many families [18–22], of which the 2,3-di-O-acyltrehalose (DAT) family was previously defined as glycolipid B. DAT is a mycobacterial factor capable of modulating host immune responses [23] and can inhibit the proliferation of murine T cells [24]. DAT along with pentaacyl trehalose (PAT) also play an important role in pathogenesis and a structural role in the cell envelope, promoting the intracellular survival of the bacterium [25]. The DATs from *M. tuberculosis* and *M. fortuitum* have been shown to be antigenic [24,26,27] and their potential use in serodiagnosis has been postulated [28,29].

Besra and coworker defined the structures of the acylated trehalose lipid family using gas chromatography-mass spectrometry, in conjunction with normal/reversed phase TLC, and one/two-dimensional ^{1}H, and ^{13}C nuclear magnetic resonance spectroscopy [21]. However, there is no report in the detailed structural assignment of the various molecular species with many isomeric structures for the entire lipid family.

In this report, we applied multiple stage precursor ion isolation and resonance excitation activation to generate distinct MS^n spectra to explore the structure details of the 2,3-diacyl trehalose (DAT) lipid family found in *M. tuberculosis*. This study highlights the unique feature of the technique of LIT MS^n for tandem-in-time precursor ion separation that affords structural characterization of a complex lipid family, while a similar structural analysis would be very difficult to perform utilizing the tandem mass spectrometric approach with a sector, TSQ or QTOF instrument.

2. Materials and Methods

2.1. Materials

All solvents (spectroscopic grade) and chemicals (ACS grade) were obtained from Sigma Chemical Co. (St. Louis, MO, USA).

2.2. Sample Preparation

M. tuberculosis strain H37Rv were grown and total lipids were extracted and isolated as previously described [8]. Briefly, the total lipid was separated by a Phenomenex C18 Kinetex (100 × 4.6 mm, pore size 100 Å, particle size 2.6 µm) column at a flow rate of 300 µL/min with a gradient system as previously described [12]. DAT (eluted at 24.45–26.43 min) fraction from three injections (~200 µg total lipid/injection) were collected and pooled, dried under a stream of nitrogen, and stored at −20 °C until use.

2.3. Mass Spectrometry

Both high-resolution (R = 100,000 at m/z 400) low-energy CID and higher collision-energy dissociation (HCD) tandem mass spectrometric experiments were conducted on a Thermo Scientific (San Jose, CA, USA) LTQ Orbitrap Velos mass spectrometer (MS) with Xcalibur operating system. Samples in CH_3OH were infused (1.5 µL/min; ~10 pmol/µL) to the ESI source, where the skimmer was set at ground potential, the electrospray needle was set at 4.0 kV, and temperature of the heated capillary was 300 °C. The automatic gain control of the ion trap was set to 5×10^4, with a maximum injection time of 100 ms. Helium was used as the buffer and collision gas at a pressure of 1×10^{-3} mbar (0.75 mTorr). The MS^n experiments were carried out with an optimized relative collision energy ranging from 25–35% and with an activation q value at 0.25. The activation time was set for 10 ms to leave a minimal residual abundance of precursor ion (around 20%). For the HCD experiments, the collision energy was set at 50–55% and mass scanned from m/z 100 to the upper m/z value that covers the precursor ions. The mass selection window for the precursor ions was set at 1 Da wide to admit the monoisotopic peak to the ion-trap for collision-induced dissociation (CID) for unit resolution detection in the ion-trap or high resolution accurate mass detection in the Orbitrap mass analyzer. Mass spectra were accumulated in the profile mode, typically for 3–10 min for MS^n spectra (n = 2,3,4).

2.4. Nomenclature

To facilitate data interpretation, the following abbreviations were adopted. The abbreviation of the long-chain fatty acid such as the stearic acid attached to the C2 position of the trehalose backbone is designated as 18:0. The multiple methyl-branched mycolipenic acid, for example, the 2,4,6-trimethyl-2-tetracosenoic acid attached to the C3-position is designated as 27:1 to reflect the fact that the acid contains a C_{27} acyl chain with one double bond. Therefore, the DAT species consisting of 18:0- and 27:1-FA substituents at C2-, and C3-position, respectively, is designated as 18:0/27:1-DAT.

3. Results and Discussion

DAT formed $[M + Alk]^+$ ions (Alk = NH_4, Li, Na, etc.) in the positive ion mode; and $[M + X]^-$ (X = Cl, RCO_2; R = H, CH_3, C_2H_5, etc.) ions in the negative-ion mode when subjected to ESI in the presence of Alk^+ and X^-. For example, when dissolved in CH_3OH with the presence of HCO_2Na, adduct ions in the fashions of $[M + Na]^+$ (Figure 1) in the positive ion mode and $[M + HCO_2]^-$ ions (data not shown) in the negative ion mode were observed. The formation of these adduct ions was revealed by the elemental composition of the molecular species deduced by high resolution mass spectrometry (Table 1). Upon being subject to CID in a linear ion-trap, the MS^n (n = 2,3,4) spectra of both the $[M + Na]^+$ and $[M + HCO_2]^-$ ions contain rich structural information readily applicable for structural identification.

Table 1. The high resolution mass measurements of the [M + Na]$^+$ ions of DATs isolated from *M. tuberculosis* and the assigned structues (* structure not defined).

Measured m/z	Theo. Mass m/z	Deviation mmu	Rel. Intensity %	Ele. Composition	Major structures	Minor structures
935.6436	935.6430	0.55	3.42	C51H92O13Na	14:1/25:1	12:1/27:1
937.6594	937.6587	0.73	4.34	C51H94O13Na	14:0/25:1	12:0/27:1; 16:1/23:0; 15:1/24:0; 13:0/26:1
939.6747	939.6743	0.37	1.48	C51H96O13Na	*	
949.6585	949.6587	−0.19	6.58	C52H94O13Na	16:1/24:1	
951.6742	951.6743	−0.15	20.8	C52H96O13Na	16:1/24:0	16:0/24:1; 15:0/25:1; 17:1/23:0; 14:0/26:1; 13:0/27:1
953.6898	953.6900	−0.12	18.49	C52H98O13Na	16:0/24:0	
961.6587	961.6587	0.01	4.31	C53H94O13Na	*	
963.6743	963.6743	−0.05	44.16	C53H96O13Na	16:1/25:1; 14:1/27:1	
965.6898	965.6900	−0.12	50.07	C53H98O13Na	16:0/25:1	14:0/27:1; 18:0/23:1
967.7054	967.7056	−0.22	12.26	C53H100O13Na	18:0/23:0; 17:0/24:0	16:0/25:0
975.6740	975.6743	−0.3	3.03	C54H96O13Na	*	
977.6899	977.6900	−0.06	25.16	C54H98O13Na	16:1/26:1	
979.7056	979.7056	−0.06	54.48	C54H100O13Na	18:0/24:1; 18:1/24:0	16:0/26:1; 17:0/25:1; 16:1/26:0; 15:0/27:1
981.7211	981.7213	−0.15	62.62	C54H102O13Na	18:0/24:0	17:0/25:0; 16:0/26:0
989.6898	989.6900	−0.17	18.03	C55H98O13Na	16:1/27:2	
991.7054	991.7056	−0.2	100	C55H100O13Na	16:1/27:1	18:1/25:1; 16:0/27:2; 17:0/26:2
993.7210	993.7213	−0.29	95.91	C55H102O13Na	16:0/27:1	18:0/25:1; 17:0/26:1
1005.7211	1005.7213	−0.2	34.31	C56H102O13Na	16:1/28:1; 17:1/27:1	17:0/27:2; 18:1/26:1; 18:0/26:2
1007.7366	1007.7369	−0.31	51.54	C56H104O13Na	17:0/27:1	16:0/28:1; 18:0/26:1; 19:0/25:1; 16:1/28:0
1009.7518	1009.7526	−0.75	7.65	C56H106O13Na	18:0/26:0	20:0/24:0
1017.7209	1017.7213	−0.38	11.79	C57H102O13Na	18:1/27:2	
1019.7367	1019.7369	−0.25	52.39	C57H104O13Na	18:1/27:1	18:0/27:2; 17:1/28:1
1021.7522	1021.7526	−0.37	92.1	C57H106O13Na	18:0/27:1	17:0/28:1
1033.7521	1033.7526	−0.47	10.49	C58H106O13Na	18:1/28:1	19:0/27:2; 19:1/27:1; 18:0/28:2; 19:0/27:2
1035.7678	1035.7682	−0.45	13.33	C58H108O13Na	19:0/27:1; 18:0/28:1	18:1/28:0; 17:0/29:1
1047.7677	1047.7682	−0.5	2.56	C59H108O13Na	18:0/28:2	
1049.7835	1049.7839	−0.31	5.47	C59H110O13Na	19:0/28:1	20:0/27:1; 18:0/29:1; 17:0/30:1
1061.7837	1061.7839	−0.19	1.63	C60H110O13Na	18:1/30:1	
1063.7991	1063.7995	−0.38	1.83	C60H112O13Na	18:0/30:1	

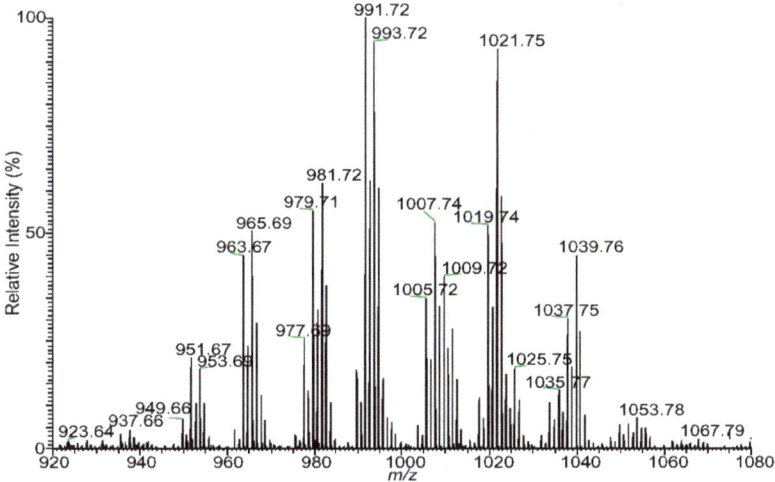

Figure 1. The positive-ion ESI mass spectrum of the [M + Na]$^+$ ions of the DAT species isolated from cell envelope of M. tuberculosis.

3.1. The Fragmentation Processes of the [M + Na]$^+$ Ions of DAT Revealed by LIT MS^n

The utility in the performance of sequential precursor ion separation, CID, and acquiring MS^n spectra of LIT MS^n mass spectrometry permits insight into not only the fragmentation processes, but also the structural details of the molecules. For example, the LIT MS^2 spectrum of the [M + Na]$^+$ ions of 18:0/27:1-DAT at m/z 1021 contained the dominated ions of m/z 859 (Figure 2a) arising from loss of glucose residue, along with the ion set at m/z 737 and 613, arising from losses of 18:0-, and 27:1-fatty acid substituents, respectively (Scheme 1). The ions of m/z 859 represent the sodiated diacylglucose with both the 18:0-, and 27:1-fatty acyl substituents. This notion is further supported by the MS^3 spectrum of the ions of m/z 859 (1021 → 859, Figure 2b) which contained ions of m/z 575 (859 − 284) and 451 (859 − 408), arising from losses of 18:0-, and 27:1-fatty acid substituents, respectively. The results also suggest that the Na$^+$ charge site most likely resides at the glucose ring with the two acyl groups (Glc 1). In contrast, the MS^2 spectrum of the [M + Na]$^+$ ions of the 6,6'-dioleoyltrehalose standard [30] at m/z 893 forms abundant ions at m/z 467, representing the sodiated oleoylglucose (data not shown), consistent with the fact that the 18:1 fatty acyl substituents in the 18:1/18:1-DAT are located on the separate Glc (i.e., Glc1 and Glc 2). Further dissociation of the ions of m/z 737 (1013 → 737, Figure 2c) gave rise to the prominent ions of m/z 329 by loss of 27:1-fatty acid substituent, and m/z 575, arising from loss of glucose residue (Glc 2), together with m/z 431 representing a sodiated ion of 27:1-FA. These results further support the fragmentation processes as proposed in Scheme 1.

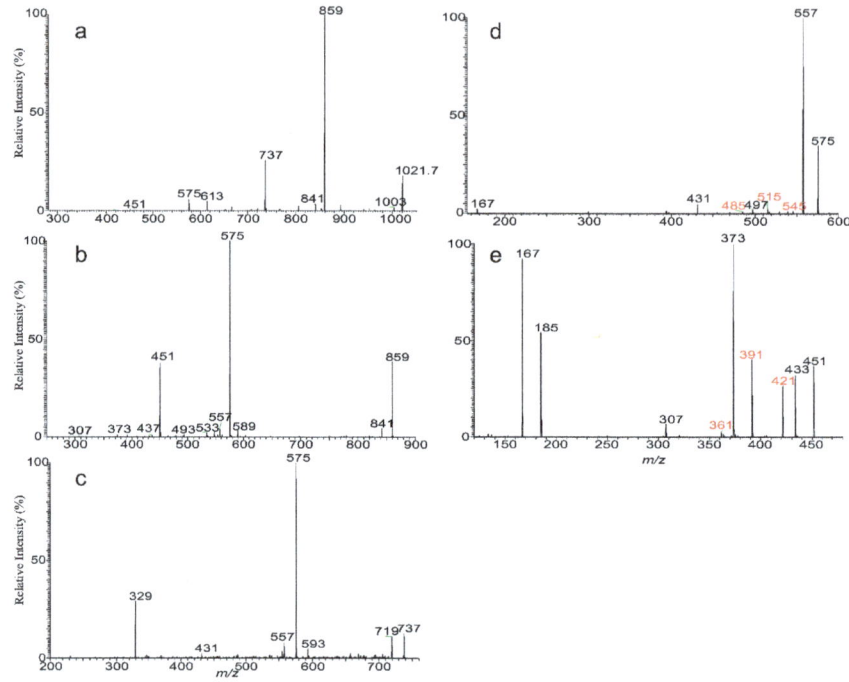

Figure 2. The LIT MS² spectrum of the [M + Na]⁺ ion of 18:0/27:1-DAT at m/z 1021 (**a**), its MS³ spectra of the ions of m/z 859 (1021 → 859) (**b**), and of m/z 737 (1021 → 737) (**c**); and its MS⁴ spectra of the ions of m/z 575 (1021 → 859 → 575) (**d**), and of m/z 451 (1021 → 859 → 451) (**e**).

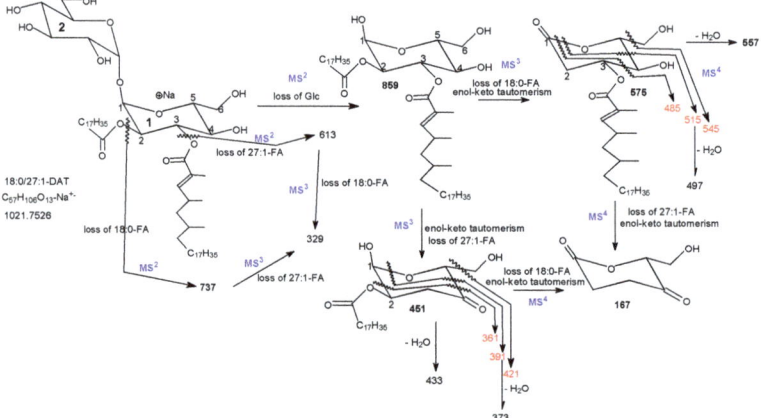

Scheme 1. The structure of [M + Na]⁺ ion of 18:0/27:1-DAT at m/z 1021 and proposed LIT MSⁿ fragmentation processes*. * All the ions represent the sodiated species. To simplify, the drawing of "Na⁺" is omitted from the scheme.

The formation of the ions of m/z 575 from m/z 859 by loss of 18:0-FA residue at C2 may involve the participation of the hydrogen atom at C1 to form an enol, which undergoes enol-keto tautomerism to yield a stable sodiated ion of monoacyl (27:1) glucose as the keto form (Scheme 1). This fragmentation processes are further supported by MS⁴ on the ions of m/z 575 (1013 → 859 → 575, Figure 2d), which yielded ions of m/z 545, 515, and 475, likely arising from the across cleavages of the glucose ring, suggesting that the 27:1-fatty acyl substituent is located at C3 (Scheme 1).

Similarly, MS^4 on the ion of m/z 451 (1013 → 859 → 451, Figure 2e) gave rise to ions of m/z 421, 391, and 361 arising from the similar rupture of the glucose ring, indicating that the 18:0-fatty acyl substituent is most likely located at C2 of the glucose ring. The preliminary loss of the 27:1-FA substituent may involve the participation of the adjacent hydrogen at C4 of Glc 1 to form an enol, which sequentially rearranges to keto form via the similar enol-keto tautomerism mechanism.

The preferential formation of the ions of m/z 575 from loss of the 18:0-FA substituent over the ions of m/z 451 from similar loss of the 27:1-FA as seen in Figure 2a is readily applicable for locating the FA substituents on the trehalose backbone.

3.2. LIT MS^n on the [M + Na]$^+$ Ions of DAT for Stereoisomer Recognition

To define the structures of DAT species with many isomeric structures using LIT MS^n is exemplified by characterization of the [M + Na]$^+$ ions of m/z 979, which gave rise to the prominent ions at m/z 817 (Figure 3a) arising from loss of glucose. Further dissociation of the ions of m/z 817 (979 → 817; Figure 3b) yielded the ion pairs of m/z 533/451, arising from losses of 18:0/24:1 fatty acid substituents, indicating that the these two acyl groups are situated at Glc 1, giving assignment of the 18:0/24:1-DAT structure. The spectrum also contained the m/z 535/449, 561/423, 547/437, 563/421, 575/409 ion pairs, arising from losses of 18:1/24:0, 16:0/26:1, 17:0/25:1, 16:1/26:0, and 15:0/27:1 FA pairs, respectively. The results indicate the presence of the 18:1/24:0-, 16:0/26:1-, 17:0/25:1-, 16:1/26:0-, and 15:0/27:1-DAT isomers. The above structure assignments were further confirmed by the MS^3 and MS^4 spectra. For example, the MS^3 spectrum of the ions of m/z 697 (979 → 697; Figure 3c) from primary loss of a 18:1-FA residue at C2 (Figure 3a) gave the abundant ions of m/z 535 (loss of Glc 2), along with ions of m/z 329 arising from loss of 24:0-FA substituent, and of m/z 391, representing a sodiated 24:0-FA cation. The results confirm the presence of 18:1/24:0-DAT. The MS^4 spectrum of the ions of m/z 535 (979 → 817 → 535; Figure 3d) contained ions of m/z 517 (loss of H_2O) and 505, 475, 445 arising from cleavages of the sugar ring similar to that shown in Scheme 1, along with ions of m/z 167 from loss of 24:0-FA, pointing to notion that the 24:0-FA is located at C3.

The MS^3 spectrum of the ions of m/z 695 (979 → 695; Figure 3e) contained the ions of m/z 533 and 329 arising from further losses of Glc and 24:1-FA residues (Figure 3a), respectively. The MS^4 spectrum of the ions of m/z 533 (979 → 817 → 533; data not shown) gave ions of m/z 503, 473, and 443 from the similar fragmentation processes that cleave the sugar ring (Scheme 1), confirming that the 24:1-FA substituent is located at C3. The results led to assign the 18:0/24:1-DAT structure. Using this LIT MS^n approach, a total of six isomeric structures were identified.

3.3. The Fragmentation Processes of the [M + HCO$_2$]$^-$ Ions of DAT Revealed by LIT MS^n

In the negative-ion mode in the presence of HCO_2^-, 18:0/27:1-DAT formed [M + HCO_2]$^-$ ions of m/z 1043, which gave rise to the prominent ions of m/z 997 by loss of HCO_2H, along with ions of m/z 713 and 589 by further losses of 18:0- and 27:1-FA substituents, respectively (Figure 4a) (Scheme 2). This fragmentation process is further supported by the MS^3 spectrum of the ions of m/z 997 (1043 → 997, Figure 4b), which are equivalent to the [M − H]$^-$ ions of 18:0/27:1-DAT. The ions of m/z 731 (Figure 4a) arising from loss of 18:0-FA as a ketene is more prominent than the ions of m/z 607 arising from analogous 27:1-ketene loss. This preferential formation of m/z 731 corresponding to loss of the FA-ketene at C2 over ions of m/z 607 from the FA-ketene loss at C3 was seen in all the MS^2 spectra of the [M + HCO_2]$^-$ ions of DAT, providing useful information for distinction of the FA substituents at C2 and C3. The ketene loss process probably involves the participation of HCO_2^-, which attracts the labile α-hydrogen on the fatty acid group to eliminate FA-ketene and HCO_2H simultaneously (Scheme 2). Therefore, the low abundance of the ions of m/z 607 arising from loss of the FA-ketene at C3 may reflect the fact that the 27:1-FA substituent at C3 contains an α-methyl side chain [19,21,22,31], and does not contain labile α-hydrogen required for ketene loss. This is in contrast to the 18:0-FA substituent at C2, which possesses two α-hydrogens (Scheme 2).

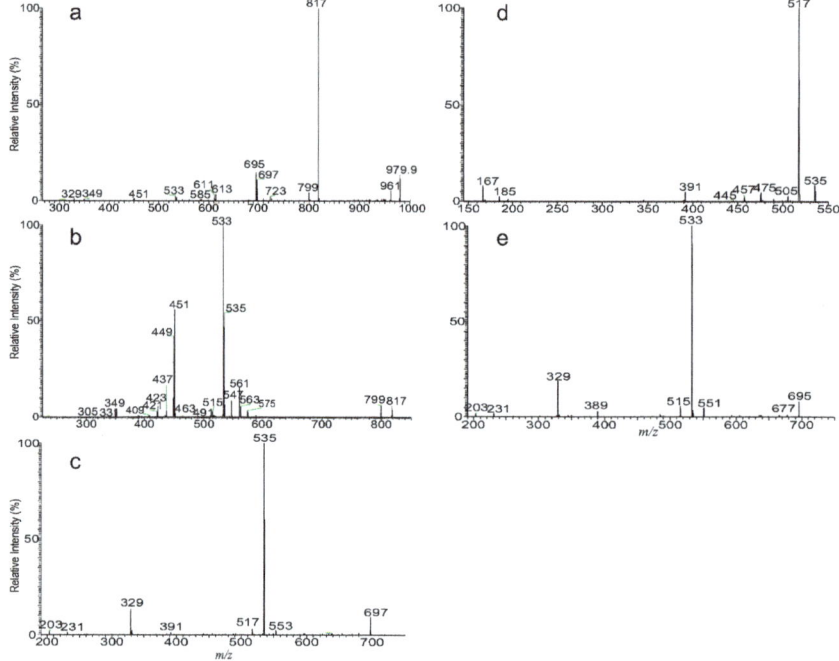

Figure 3. The MS^2 spectrum of the [M + Na]$^+$ ions of m/z 979 (**a**), its MS^3 spectra of the ions of m/z 817 (979 → 817)) (**b**), of m/z 697 (979 → 697) (**c**), its MS^4 spectrum of the ions of m/z 535 (979 → 817 → 535) (**d**); and the MS^3 spectrum of the ions of m/z 695 (979 → 695) (**e**).

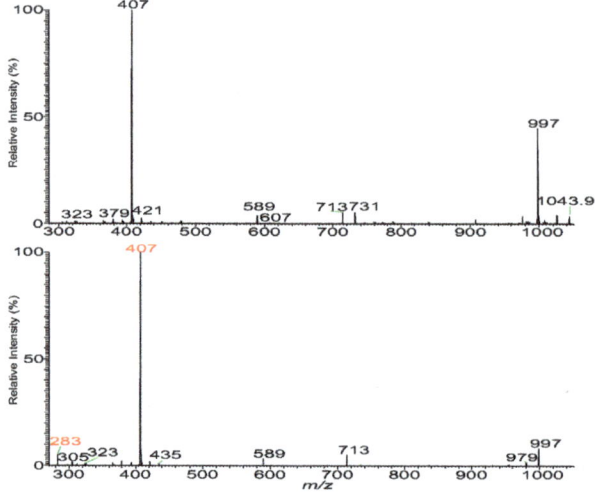

Figure 4. The negative-ion MS^2 spectrum of the [M + HCO$_2$]$^-$ ions of 18:0/27:1-DAT at m/z 1043 (**a**), its MS^3 spectrum of the ions of m/z 997 (1043 → 997) (**b**).

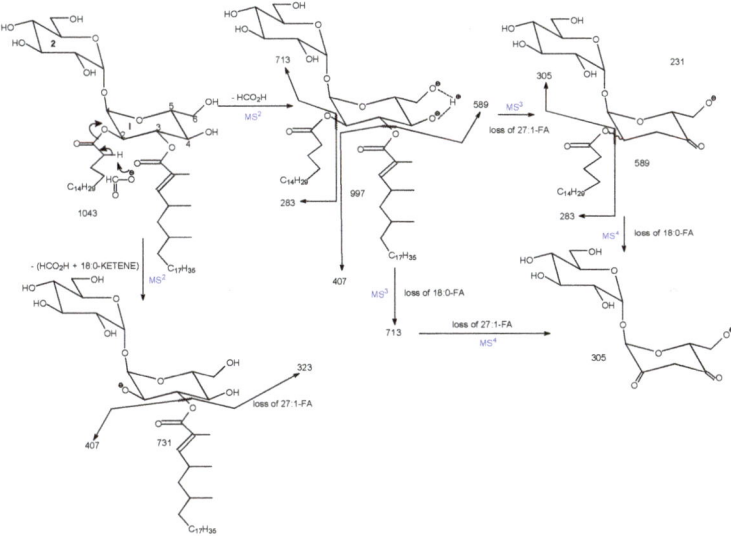

Scheme 2. The proposed LIT MS^n fragmentation processes of the [M + HCO$_2$]$^-$ ions of 18:0/27:1-DAT at m/z 1043.

The ions at m/z 731 and 607 arising from losses of 18:0-ketene and 27:1-ketene, respectively, are absent in Figure 4b. This is consistent with the notion that the ketene loss requires the participation of HCO$_2$$^-$. The ketene loss pathway becomes not operative after the [M − H]$^-$ ions are formed from [M + HCO$_2$]$^-$ by loss of HCO$_2$H.

The spectrum (Figure 4b) also contained the prominent ions of m/z 407, representing 27:1-fatty acid carboxylic anions, and the ions of m/z 283 representing 18:0-FA carboxylate anions, along with ions of m/z 305 arising from losses of both 18:0- and 27:1-FA substituents. The preferential formation of the ions of m/z 407 (at C3) over m/z 283 (at C2) is also a reflection of the location of the fatty acid substituents on the Glc ring, leading to the assignment of 18:0/27:1-DAT structure.

3.4. Recognition of Stereoisomers Applying LIT MS^n on the [M + HCO$_2$]$^-$ Ions

Similarly, the MS^2 spectrum of the ions of m/z 1001 is dominated by the ions of m/z 955 from loss of HCO$_2$H (Figure 5a). The MS^3 spectrum of the ions of m/z 955 (1001 → 955) (Figure 5b) contained ions at m/z 701, 699, 685, 673, 671, 589, 587, 575, 561, 559, similar to those seen in Figure 5a, consistent with the consecutive dissociation processes of the [M − H]$^-$ ions that eliminate the FA substituents. These ions arose from losses of 16:1-, 16:0-, 17:0-, 18:1-, 18:0-, 24:1-, 24:0-, 25:0-, 26:1, and 26:0-fatty acid substituents, respectively. The spectrum also contained the major m/z 283/365, 281/367 ion pairs, together with the minor ion pairs of m/z 255/393, 253/395, 269/379, 241/407. These structural information led to assignment of the major 18:0/24:1-, and 18:1/24:0-DAT isomers, together with the 16:0/26:1-, 16:1/26:0-, 17:0/25:1, and 15:0/27:1-DAT minor isomers. It should be noted that the absence of ions at m/z 241, 253, 255, and 269 representing 15:0-, 16:1-, 16:0-, and 17:0-carboxylate anions, respectively, in Figure 5b, are attributable to the low mass cutoff nature of an ion-trap instrument. In contrast, these ions are abundant in the HCD production ion spectrum (data not shown), similar to that obtained by a triple quadrupole instrument [32]. The location of the fatty acid substituent position on the glucose skeleton is again confirmed by observation of the ions corresponding to loss of the fatty acid substituents as ketenes. For example, ions at m/z 691 and 689 (Figure 5a) derived from losses of (HCO$_2$H + 18:1-ketene) and (HCO$_2$H + 18:0-ketene), respectively, pointing to the notion that both 18:1- and 18:0-FA are situated at C2 of the 18:1/24:0-, and 18:0/24:1-DAT isomers, respectively. The assigned structure of 18:0/24:1-DAT, for example, is further confirmed by MS^4 on the ions of m/z 671 (1001 → 955 → 671) (Figure 5c), which formed ions of m/z 365, representing a 24:1-carboxylate anion, and ions of m/z 323 and 305 arising from further loss of 24:1-FA substituent as ketene and FA, respectively.

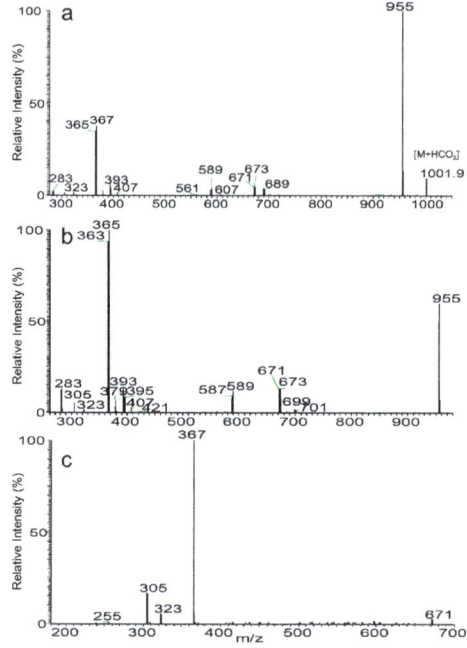

Figure 5. The negative-ion MS^2 spectrum of the [M + HCO$_2$]$^-$ ions of m/z 1001 (**a**), its MS^3 spectrum of the ions of m/z 955 (1001 → 955) (**b**), and the MS^4 spectrum of the ions of m/z 671 (1001 → 955 → 671) (**c**).

3.5. Characterization of Minor Species Applying LIT MS^n on the [M + HCO$_2$]$^-$ Ions

Applying multiple-stage mass spectrometry (LIT MS^n) for consecutive ion separation followed by CID mass spectrometry is particularly useful for characterization of minor DAT species as [M + HCO$_2$]$^-$ ions. For example, the MS^2 spectrum of the [M + HCO$_2$]$^-$ ion of the minor DAT at m/z 989 (Figure 6a) gave a major [M − H]$^-$ fragment ions at m/z 943, but the spectrum also contained many unrelated fragment ions (e.g., ions of m/z 957, 930, 921, and 905) that complicate the structural identification. These fragment ions may arise from the adjacent precursor ions admitted together with the desired DAT ions for CID, due to that the precursor ion selection window (1 Da) cannot sufficiently isolate the isobaric ions (the mass selection window and injection time govern the total ions admitted to the trap for CID and >1 Da mass selection window is often required to maintain the sensitivity). Thus, fragment ions unrelated to the targeted molecule were formed simultaneously and complicating the structure analysis. However, the MS^3 spectrum of m/z 943 (Figure 6b) contained only the fragment ions related to the DAT species, due to that the [M − H]$^-$ ions still retain the complete structure but have been further segregated, and fragment ions unrelated to the structure have been filtrated by another stage (MS^3) isolation. In this context, the MS^4 spectrum of the ions of m/z 589 (989 → 943 → 589; Figure 6c), which were further "purified", becomes even more specific, due to that only the fragment ions from DAT that consists of 23:0-FA substituent at C3 were subjected to further CID. Thus, the spectrum only contained ions of m/z 283, representing a 18:0-carboxylate anion, along with ions at m/z 323 and 305, representing the dehydrated trehalose anions. These results led to specifically define the 18:0/23:0-DAT structure. The spectrum (Figure 6b) also contained the m/z 269/367 and 255/381 ion pairs, representing the 17:0/24:0-, 16:0/25:0-FA carboxylate anion pairs, together with ions of m/z 687/561 and 673/575 ion pairs, arising from losses of 17:0/24:0-, 16:0/25:0-FA substituents, respectively. These results readily led to the assignment of 17:0/24:0- and 16:0/25:0-DAT isomeric structures.

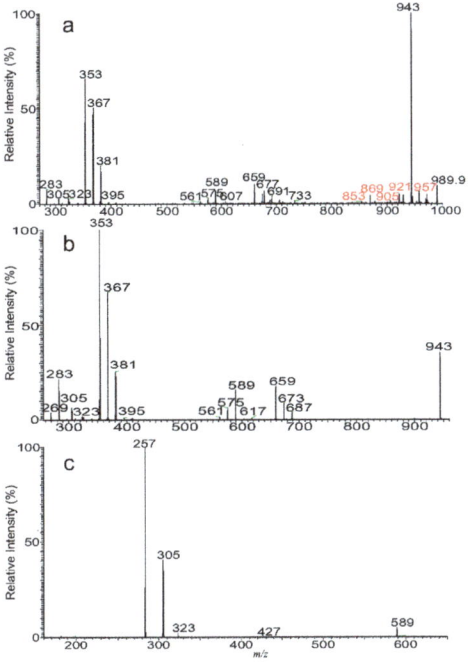

Figure 6. The negative-ion MS^2 spectrum of the [M + HCO$_2$]$^-$ ions of DAT at m/z 989 (**a**), its MS^3 spectrum of the ions of m/z 943 (989 → 943) (**b**), MS^4 spectrum of the ions of m/z 589 (989 → 943 → 589) (**c**). The ions of m/z 989 are minor species and its MS^2 spectrum contains several ions (shown in red in (a)) unrelated to the structure, but have been filtrated from higher stage MS^n, as shown in (b) (MS^3) and (c) (MS^4).

4. Conclusions

Sequential precursor ion isolation applying multiple-stage mass spectrometry (LIT MS^n) adds another dimension of separation in the analysis, providing a powerful tool for structural identification of various compounds. Thereby, many isomeric structures of the molecule can be unveiled in a very short period of time. By contrast, using the conventional chromatographic separation combined with a TSQ or QTOF instrument, the consecutive precursor ion isolation by MS is not achievable, and the species separation can only rely on column separation. Thus, complete separation of a complex lipid mixture with a wide range of molecular species and many isomeric structures is often difficult. Compound separation by chromatographic means also requires significantly more times [33], as compared to the LIT MS^n approach, by which the separation-CID-detection process can be completed within a very short period of time.

LIT MS^n permits ion isolation in the time sequence manner, and the separation of ions is flexible (i.e., the types of ions selected and the mass selection window of precursor ions). The selected ions become more specific, and the MS^n spectrum provides more structurally specific information as the MS^n stage advances, therefore, resulting in a confident and detailed structural identification. The structures of minute ion species that are often difficult to define by other analytical method can also be assigned (Table 1). However, the sensitivity declines as the higher order of MS^n stage proceeds.

Other drawback includes that a complete structural information is not necessary extractable by MS^n. For example, the positions of the double bond and methyl side chain of the fatty acid substituents at C3 have not been defined in this study.

The LIT MS^n approach as described here affords near complete structural characterization of a complex DAT lipid family, locating the fatty acyl groups on the trehalose backbone, and recognizing many isomeric structures. A LIT MS^n approach combined with chemical reaction modification [9,34] for locating the functional groups including the methyl, hydroxyl, and the double bond on the fatty acid substituents are currently in progress in our laboratory.

Author Contributions: C.F. performs lipid separation; R.B.A. and G.E.P. grow cell and extract total lipids; L.L. and L.L. synthesize DAT standards. F.-F.H. performs mass spectrometry analysis and write the manuscript.

Funding: This research was funded by US Public Health Service Grants P41GM103422, P30DK020579, R01AI130454, R21AI128427, R21HL120760, and R21AI113074.

Acknowledgments: This research was supported by US Public Health Service Grants P41GM103422, P30DK020579, R01AI130454, R21AI128427, R21AI113074 and R21HL120760.

Conflicts of Interest: The authors declare no conflict of interest.

Abbreviations

LIT, linear ion trap; MS^n, multiple stage mass spectrometry; Glc, glucose; DAT, diacyltrehalose; FA, fatty acid; FA-ketene, fatty acyl ketene; CID, collision induced dissociation; ESI, electrospray ionization; HRMS, high-resolution mass spectrometry.

References

1. Johnson, J.V.; Yost, R.A.; Kelley, P.E.; Bradford, D.C. Tandem-in-space and tandem-in-time mass spectrometry: triple quadrupoles and quadrupole ion traps. *Anal. Chem.* **1990**, *62*, 2162–2172. [CrossRef]
2. Markarov, A.; Cousijn, E.; Cantebury, J.; Denisov, E.; Thoeing, C.; Lange, O.; Kreutzman, A.; Ayzikov, K.; Damoc, E.; Tabiwang, A.; et al. Extension of Orbitrap capabilities to enable new applications. In Proceedings of the 65th Conference on Mass Spectrometry and Allied Topics, Indianapolis, IN, USA, 4 June 2017.
3. Vogel, C.; Marcotte, E.M. Insights into the regulation of protein abundance from proteomic and transcriptomic analyses. *Nat. Rev. Genet.* **2012**, *13*, 227–232. [CrossRef] [PubMed]
4. Perry, R.H.; Cooks, R.G.; Noll, R.J. Orbitrap mass spectrometry: Instrumentation, ion motion and applications. *Mass Spectrom. Rev.* **2008**, *27*, 661–699. [CrossRef] [PubMed]
5. Senyuva, H.Z.; Gökmen, V.; Sarikaya, E.A. Future perspectives in Orbitrap™-high-resolution mass spectrometry in food analysis: A review. *Food Addit. Contam. A* **2015**, *32*, 1568–1606. [CrossRef] [PubMed]
6. Jabbour, R.E.; Snyder, A.P. Chap. 14—Mass spectrometry-based proteomics techniques for biological identification. In *Biological Identification*; Schaudies, R.P., Ed.; Woodhead Publishing: Sawston, UK, 2014; pp. 370–430.
7. Eliuk, S.; Makarov, A. Evolution of Orbitrap Mass Spectrometry Instrumentation. *Ann. Rev. Anal. Chem.* **2015**, *8*, 61–80. [CrossRef] [PubMed]
8. Rhoades, E.R.; Streeter, C.; Turk, J.; Hsu, F.-F. Characterization of Sulfolipids of Mycobacterium tuberculosis H37Rv by Multiple-Stage Linear Ion-Trap High-Resolution Mass Spectrometry with Electrospray Ionization Reveals That the Family of Sulfolipid II Predominates. *Biochemistry* **2011**, *50*, 9135–9147. [CrossRef] [PubMed]
9. Hsu, F.-F. Characterization of Hydroxyphthioceranoic and Phthioceranoic Acids by Charge-Switch Derivatization and CID Tandem Mass Spectrometry. *J. Am. Soc. Mass Spectrom.* **2016**, *27*, 622–632. [CrossRef]
10. Hsu, F.F.; Turk, J.; Owens, R.M.; Rhoades, E.R.; Russell, D.G. Structural Characterization of Phosphatidyl-myo-Inositol Mannosides from Mycobacterium bovis Bacillus Calmette Guerin by Multiple-Stage Quadrupole Ion-Trap Mass Spectrometry with Electrospray Ionization. II. Monoacyl- and Diacyl-PIMs. *J. Am. Soc. Mass Spectrom.* **2007**, *18*, 479–492. [CrossRef]
11. Hsu, F.F.; Turk, J.; Owens, R.M.; Rhoades, E.R.; Russell, D.G. Structural characterization of phosphatidyl-myo-inositol mannosides from Mycobacterium bovis Bacillus Calmette Guerin by multiple-stage quadrupole ion-trap mass spectrometry with electrospray ionization. I. PIMs and lyso-PIMs. *J. Am. Soc. Mass Spectrom.* **2007**, *18*, 466–478. [CrossRef]

12. Flentie, K.N.; Stallings, C.L.; Turk, J.; Minnaard, A.J.; Hsu, F.-F. Characterization of phthiocerol and phthiodiolone dimycocerosate esters of M. tuberculosis by multiple-stage linear ion-trap MS. *J. Lipid Res.* **2016**, *57*, 142–155. [CrossRef]
13. Hoppe, H.C.; de Wet, B.J.; Cywes, C.; Daffe, M.; Ehlers, M.R. Identification of phosphatidylinositol mannoside as a mycobacterial adhesin mediating both direct and opsonic binding to nonphagocytic mammalian cells. *Infect. Immun.* **1997**, *65*, 3896–3905. [PubMed]
14. Howard, N.C.; Marin, N.D.; Ahmed, M.; Rosa, B.A.; Martin, J.; Bambouskova, M.; Sergushichev, A.; Loginicheva, E.; Kurepina, N.; Rangel-Moreno, J.; et al. Mycobacterium tuberculosis carrying a rifampicin drug resistance mutation reprograms macrophage metabolism through cell wall lipid changes. *Nat. Microbiol.* **2018**, *3*, 1099–1108. [CrossRef] [PubMed]
15. DaffÉ, M.; Lacave, C.; LanÉElle, M.-A.; Gillois, M.; LanÉElle, G. Polyphthienoyl trehalose, glycolipids specific for virulent strains of the tubercle bacillus. *Eur. J. Biochem.* **1988**, *172*, 579–584. [CrossRef] [PubMed]
16. Minnikin, D.E.; Dobson, G.; Sesardic, D.; Ridell, M. Mycolipenates and Mycolipanolates of Trehalose from Mycobacterium tuberculosis. *J. Gen. Microbiol.* **1985**, *131*, 1369–1374. [CrossRef] [PubMed]
17. Munoz, M.; Laneelle, M.A.; Luquin, M.; Torrelles, J.; Julian, E.; Ausina, V.; Daffe, M. Occurrence of an antigenic triacyl trehalose in clinical isolates and reference strains of Mycobacterium tuberculosis. *FEMS Microbiol. Lett.* **1997**, *157*, 251–259. [CrossRef]
18. Lemassu, A.; Laneelle, M.A.; Daffe, M. Revised structure of a trehalose-containing immunoreactive glycolipid of Mycobacterium tuberculosis. *FEMS Microbiol. Lett.* **1991**, *62*, 171–175. [CrossRef] [PubMed]
19. Ariza, M.A.; Martín-Luengo, F.; Valero-Guillén, P.L. A family of diacyltrehaloses isolated from Mycobacterium fortuitum. *Microbiology* **1994**, *140*, 1989–1994. [CrossRef] [PubMed]
20. Ariza, M.A.; Valero-Guillen, P.L. Delineation of molecular species of a family of diacyltrehaloses from Mycobacterium fortuitum by mass spectrometry. *FEMS Microbiol. Lett.* **1994**, *119*, 279–282. [CrossRef]
21. Besra, G.S.; Bolton, R.C.; McNeil, M.R.; Ridell, M.; Simpson, K.E.; Glushka, J.; Van Halbeek, H.; Brennan, P.J.; Minnikin, D.E. Structural elucidation of a novel family of acyltrehaloses from Mycobacterium tuberculosis. *Biochemistry* **1992**, *31*, 9832–9837. [CrossRef]
22. Gautier, N.; Marín, L.M.L.; Lanéelle, M.A.; Daffé, M. Structure of mycoside F, a family of trehalose-containing glycolipids of Mycobacterium fortuitum. *FEMS Microbiol. Lett.* **1992**, *77*, 81–87.
23. Lee, K.-S.; Dubey, V.S.; Kolattukudy, P.E.; Song, C.-H.; Shin, A.R.; Jung, S.-B.; Yang, C.-S.; Kim, S.-Y.; Jo, E.-K.; Park, J.-K.; et al. Diacyltrehalose of Mycobacterium tuberculosis inhibits lipopolysaccharide- and mycobacteria-induced proinflammatory cytokine production in human monocytic cells. *FEMS Microbiol. Lett.* **2007**, *267*, 121–128. [CrossRef] [PubMed]
24. Saavedra, R.; Segura, E.; Leyva, R.; Esparza, L.A.; López-Marín, L.M. Mycobacterial Di-O-Acyl-Trehalose Inhibits Mitogen- and Antigen-Induced Proliferation of Murine T Cells In Vitro. *Clin. Diagn. Lab. Immun.* **2001**, *8*, 1081–1088. [CrossRef]
25. Bailo, R.; Bhatt, A.; Ainsa, J.A. Lipid transport in Mycobacterium tuberculosis and its implications in virulence and drug development. *Biochem. Pharmacol.* **2015**, *96*, 159–167. [CrossRef] [PubMed]
26. Papa, F.; Cruaud, P.; David, H.L. Antigenicity and specificity of selected glycolipid fractions from Mycobacterium tuberculosis. *Res. Microbiol.* **1989**, *140*, 569–578. [CrossRef]
27. Hamid, M.E.; Fraser, J.L.; Wallace, P.A.; Besra, G.; Goodfellow, M.; Minnikin, D.E. Antigenic glycolipids of Mycobacterium fortuitum based on trehalose acylated with 2-methyloctadec-2-enoic acid. *Lett. Appl. Microbiol. Rev.* **1993**, *16*, 132–135. [CrossRef]
28. Ridell, M.; Wallerstr6m, G.; Minnikin, D.E.; Bolton, R.C.; Magnusson, M. A comparative serological study of antigenic glycolipids from Mycobacteriurn tuberculosis. *Tubercle Lung Dis.* **1992**, *73*, 71–75. [CrossRef]
29. Tórtola, M.T.; Lanéelle, M.A.; Martín-Casabona, N. Comparison of two 2,3-diacyl trehalose antigens from Mycobacterium tuberculosis and Mycobacterium fortuitum for serology in tuberculosis patients. *Clin. Diagn. Lab. Immun.* **1996**, *3*, 563–566.
30. Prabhakar, S.; Vivès, T.; Ferrières, V.; Benvegnu, T.; Legentil, L.; Lemiègre, L. A fully enzymatic esterification/transesterification sequence for the preparation of symmetrical and unsymmetrical trehalose diacyl conjugates. *Green Chem.* **2017**, *19*, 987–995. [CrossRef]
31. Botté, C.Y.; Deligny, M.; Roccia, A.; Bonneau, A.-L.; Saïdani, N.; Hardré, H.; Aci, S.; Yamaryo-Botté, Y.; Jouhet, J.; Dubots, E.; et al. Chemical inhibitors of monogalactosyldiacylglycerol synthases in Arabidopsis thaliana. *Nat. Chem. Biol.* **2011**, *7*, 834–842. [CrossRef]

32. Olsen, J.V.; Macek, B.; Lange, O.; Makarov, A.; Horning, S.; Mann, M. Higher-energy C-trap dissociation for peptide modification analysis. *Nat. Meth.* **2007**, *4*, 709–712. [CrossRef]
33. Hsu, F.F. Mass spectrometry-based shotgun lipidomics—A critical review from the technical point of view. *Anal. Bioanal. Chem.* **2018**, *410*, 6387–6409. [CrossRef] [PubMed]
34. Frankfater, C.; Jiang, X.; Hsu, F.F. Characterization of Long-Chain Fatty Acid as N-(4-Aminomethylphenyl) Pyridinium Derivative by MALDI LIFT-TOF/TOF Mass Spectrometry. *J. Am. Soc. Mass Spectrom.* **2018**, *29*, 1688–1699. [CrossRef] [PubMed]

© 2019 by the authors. Licensee MDPI, Basel, Switzerland. This article is an open access article distributed under the terms and conditions of the Creative Commons Attribution (CC BY) license (http://creativecommons.org/licenses/by/4.0/).

Article

Hydrophilic Monomethyl Auristatin E Derivatives as Novel Candidates for the Design of Antibody-Drug Conjugates

Filip S. Ekholm [1,†], Suvi-Katriina Ruokonen [1,†], Marina Redón [1], Virve Pitkänen [2], Anja Vilkman [2], Juhani Saarinen [2], Jari Helin [2], Tero Satomaa [2,*] and Susanne K. Wiedmer [1,*]

- [1] Department of Chemistry, University of Helsinki, PO Box 55, A. I. Virtasen aukio 1, FI 00014 Helsinki, Finland; filip.ekholm@helsinki.fi (F.S.E.); suvi.k.ruokonen@gmail.com (S.-K.R.); marina.redonmunoz@gmail.com (M.R.)
- [2] Glykos Finland Ltd., Viikinkaari 6, 00790 Helsinki, Finland; virve.pitkanen@glykos.fi (V.P.); anja.vilkman@glykos.fi (A.V.); juhani.saarinen@glykos.fi (J.S.); jari.helin@glykos.fi (J.H.)
- * Correspondence: tero.satomaa@glykos.fi (T.S.); susanne.wiedmer@helsinki.fi (S.K.W.); Tel.: +358-9-3193-6340 (T.S.); +358-405-826-629 (S.K.W.)
- † These authors contributed equally to this work.

Received: 28 October 2018; Accepted: 14 December 2018; Published: 24 December 2018

Abstract: Antibody-drug conjugates (ADCs) are promising state-of-the-art biopharmaceutical drugs for selective drug-delivery applications and the treatment of diseases such as cancer. The idea behind the ADC technology is remarkable as it combines the highly selective targeting capacity of monoclonal antibodies with the cancer-killing ability of potent cytotoxic agents. The continuous development of improved ADCs requires systematic studies on the nature and effects of warhead modification. Recently, we focused on the hydrophilic modification of monomethyl auristatin E (MMAE), the most widely used cytotoxic agent in current clinical trial ADCs. Herein, we report on the use of micellar electrokinetic chromatography (MEKC) for studying the hydrophobic character of modified MMAE derivatives. Our data reveal a connection between the hydrophobicity of the modified warheads as free molecules and their cytotoxic activity. In addition, MMAE-trastuzumab ADCs were constructed and evaluated in preliminary cytotoxic assays.

Keywords: antibody-drug conjugate; biopharmaceutical; cytotoxicity; hydrophobicity; micellar electrokinetic chromatography

1. Introduction

Natural science has witnessed many breakthroughs during the past decades. The new biological insights gained, the improved biochemical protocols and analytical tools developed, and the constant advances in chemical reaction technologies have paved the way for exciting and multidisciplinary research fields such as the field of antibody-drug conjugates (ADCs). ADCs are modern drug-delivery molecules that combine the selective targeting capabilities of monoclonal antibodies (mab) with the potent cytotoxicity displayed by toxic organic compounds [1,2]. The interest toward ADCs and the investments in ADC research have increased exponentially in recent years as a result of the U.S. Food and Drug Administration (FDA) approval of brentuximab vedotin (Adcetris®) in 2011 (for relapsed cases of Hodgkin's lymphoma and anaplastic large cell lymphoma) [3], trastuzumab emtansine (Kadcyla®) in 2013 (for human epidermal growth factor receptor 2 -positive metastatic breast cancer) [4], gemtuzumab ozogamicin (Mylotarg®) in 2017 (for acute myeloid leukemia) [5], and inotuzumab ozogamicin (Besponsa®) in 2017 (for acute lymphoblastic leukemia) [6].

The development of modern ADCs requires systematic research on antibodies, bioconjugation technologies, as well as information on the properties of the payload molecules and the characteristics of the end products. Recently, hydrophilic derivatization of payload molecules was reported to have beneficial effects on the overall properties of the ADCs, for example, on the therapeutic index and the pharmacokinetics [7,8]. In line with the current research trends, we developed an alternative strategy for increasing the hydrophilicity of the cytotoxic agents based on the incorporation of carbohydrates and we constructed a limited set of monomethyl auristatin E (MMAE)-carbohydrate hybrids [9].

MMAE (see the structure displayed in Figure 1) is an antineoplastic and antimitotic drug that appears as the cytotoxic agent in at least sixteen ADCs which have progressed to clinical trials [10,11]. Among these is the ADC brentuximab vedotin, which is used in the treatment of relapsed cases of Hodgkin's lymphoma and anaplastic large cell lymphoma [3]. On a more general level, MMAE and other auristatins have become important cytotoxic agents for ADC development since they tolerate covalent structural modifications without substantial loss of cytotoxic activity. While this is beneficial, the hydrophobic nature of MMAE and especially current MMAE-linker conjugates is sub-optimal for the development of ADCs with high drug-to-antibody ratios. This is because the attachment of multiple drug-linker moieties of this kind may lead to devastating effects on the biocompatibility and pharmaceutical efficacy of the end products. These problems are reflected in the design of current ADCs where focus is placed on low drug-to-antibody ratios, typically in the range of 2–4. In addition, multi-drug resistant cancer cells tend to overexpress efflux pump proteins capable of removing hydrophobic cytotoxic agents from the intracellular environment, thus further diminishing their potential [12]. The incorporation of hydrophilic moieties in the cytotoxic agents or the payload molecules has been identified as a valid strategy for overcoming these challenges and circumventing issues related to unwanted aggregation and clearance of ADCs [13]. To date, hydrophilic linkers based on polyethylene glycol (PEG) [7], the sulfonate group [8], and carbohydrates [9] have been reported. We have previously focused on the inclusion of hydrophilic carbohydrates in the linker species and the cytotoxic agent due to their biocompatibility, pre-existing degradation routes, further derivatization possibilities, and, just as important, low cost.

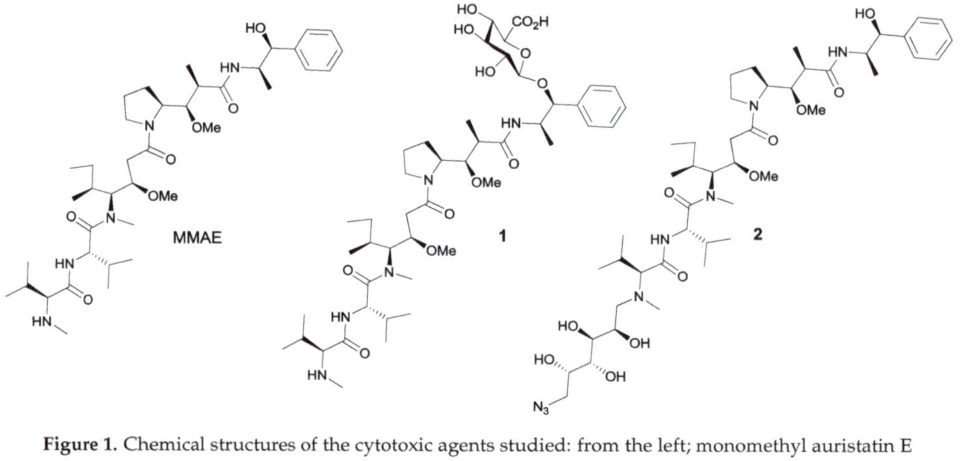

Figure 1. Chemical structures of the cytotoxic agents studied: from the left; monomethyl auristatin E (MMAE), 1 (β-d-glucuronyl-monomethylauristatin E, MMAU), and 2 (MMAE-glycolinker-substrate).

A thorough investigation of the change in the hydrophobic character was not included in the previous studies, even though it is important in understanding the nature of the modified molecules and the effects of the chosen strategy. Therefore, we continue our studies in this work by determining the relative hydrophobicities of MMAE and representatives of our own modified auristatins, namely, β-d-glucuronyl-monomethylauristatin E (MMAU, compound 1 in Figure 1) and an MMAE-glycolinker-substrate (compound 2 in Figure 1), by micellar electrokinetic chromatography (MEKC) using sodium dodecyl sulfate and sodium cholate as surfactants. Furthermore, cytotoxic assays reveal an obvious connection between the hydrophobic properties of the warhead molecules and their corresponding cytotoxicity. Due to the mechanism by which ADCs function (internalization followed by the release of the cytotoxic agent), it was important to analyze the cytotoxic profiles of eventual end products in addition to those of the free warhead molecules. As a result, trastuzumab-auristatin derivatives, which can be used to treat HER2-positive breast cancer patients, were constructed and their cytotoxicities were screened.

2. Materials and Methods

2.1. Chemicals

Reagents and solvents were purchased from commercial sources. Reactions solvents were dried and distilled prior to use when necessary. All reactions containing moisture- and/or air-sensitive reagents were carried out under argon atmosphere.

2.2. Capillary Electrophoresis

A Hewlett Packard ^{3D}CE (Agilent, Waldbronn, Germany) instrument was used for all capillary electrophoresis (CE) runs. Uncoated fused silica capillary (length 30/38.5 cm) was obtained from Polymicro Technologies (Phoenix, AZ, USA) and the inner and outer diameters of the capillary were 50 µm and 375 µm, respectively. The separation voltage was 25 kV and the capillary cassette temperature was kept constant at 25 °C. Samples were injected at 10 mbar for 10 s. Thiourea (0.5 or 0.2 mM) was used as an electroosmotic flow (EOF) marker and 10 mM (ionic strength) sodium phosphate buffer at pH 7.4 was used as the background electrolyte (BGE) solution for CE runs. In MEKC studies, the surfactants were dispersed in the same BGE solution. New capillaries were preconditioned by rinsing for 15 min with 0.1 M sodium hydroxide, 15 min with water, and for 2–5 min with the CE or MEKC BGE solution. All runs were repeated at least five times.

2.3. Calculations of Retention Factors and Distribution Constants

The retention factor (k) in chromatography is a description of the time the sample component resides in the stationary phase (or pseudostationary phase, PSP) relative to the time it resides in the mobile phase. The expression states how much longer a sample component is retarded by the stationary phase than it would take to travel through the column or capillary with the velocity of the mobile phase. In the case of CE techniques, the stationary phase is either stationary as in capillary electrochromatography or pseudostationary as in electrokinetic capillary chromatography. The velocity of the mobile phase is based on the velocity of the EOF; however, external pressure assistance is also possible. In this work we used the EKC mode and here the retention factor describes the molar ratio of an analyte in a PSP and in an aqueous mobile phase, that is, ($\frac{n_{PSP}}{n_{aq}}$). From this expression it is obvious that the retention factor in EKC is dependent on the PSP concentration. In MEKC the retention factor can be calculated using Equation 1, when the effective electrophoretic mobility of the analyte under MEKC (μ_{MEKC}) and CE conditions (μ_0) is known, as well as the effective electrophoretic mobility of the micelles (μ_{PSP}):

$$k = \frac{u_{MEKC} - u_0}{u_{PSP} - u_{MEKC}}. \tag{1}$$

An iteration procedure employing a homologous series of alkylbenzoates was used for estimating the μ_{PSP}, as previously reported [14,15]. The resulting values for 20 mM sodium dodecylsulfate (SDS) and for 20 mM SDS mixed with 40 mM sodium cholate (SC) (hereafter called 20/40 mM SDS/SC) were $-4.32\text{E-}08$ and $-4.06\text{E-}08$ m$^2 \cdot$V^{-1}s^{-1}, respectively.

The distribution constant (K_D) is the molar concentration ratio of an analyte between a pseudostationary phase and an aqueous phase and it can be calculated for systems with known phase ratios (ϕ) using Equation (2):

$$K_D = \frac{k}{\phi} \qquad (2)$$

The phase ratio is the volume ratio of the pseudostationary phase and the aqueous phase in the fused silica capillary and it can be calculated from Equation (3):

$$\phi = \frac{V_{PSP}}{V_{aq}} = \frac{v_{spec,\,vol} \cdot M \cdot (c_{PSP} - CMC)}{1 - \left(v_{spec,\,vol} \cdot M \cdot (c_{PSP} - CMC)\right)}, \qquad (3)$$

where V_{PSP} and V_{aq} are the volumes of the pseudostationary phase and the aqueous phase in the capillary, respectively, $v_{spec,vol}$ is the partial specific volume, M is the molar mass, C_{PSP} is the total concentration of the surfactants, and CMC is the critical micelle concentration of the surfactants. The partial specific volumes of SDS and SC in phosphate buffer at pH 7.4 (I = 10 mM) were approximated to be close to the value in water at 25 °C and therefore values of 0.853 mL·g^{-1} and 0.749 mL·g^{-1}, respectively, were used [16,17]. The partial specific volumes of the SDS/SC mixed micelles were estimated based on the used surfactant concentration ratios.

2.4. Determination of the Critical Micelle Concentration

For the phase ratio calculation, the surfactant CMCs in sodium phosphate buffer (pH 7.4, I = 10 mM) were determined with an optical contact angle meter using the pendant drop method (CAM 200 Optical Contact Angle Meter, Biolin Scientific, Espoo, Finland). Surface tensions of the surfactant solutions at different concentrations were determined by taking images of the pendant drops (4 drops/concentration, 20 frames/drop) with a CCD Video camera module and fitting a Young–Laplace equation to the frames using an Attension Theta Software (ver. 4.1.0. Biolin Scientific, Espoo, Finland). All measurements were repeated three times. The resulting CMCs for SDS and for the 20/40 SDS/SC mixture were 5.08 ± 0.24 and 6.00 ± 0.38 mM, respectively.

2.5. Preparation of Drug-Linker Compounds and Antibody-Drug Conjugates

MMAU and ε-maleimidocaproyl-L-valine-L-citrulline-paraaminobenzyloxycarbonyl-paranitrophenyl (MC-Val-Cit-PABC-pNP) were prepared as described previously [18]. Interchain disulphide bridges of trastuzumab (Herceptin®, Roche, Espoo, Finland) were reduced with tris(2-carboxyethyl)phosphine (TCEP) and antibody drug-conjugates (ADCs) were synthesized by incubating 0.1 mM TCEP-reduced antibody with 50× molar excess of either MC-Val-Cit-PABC-MMAU or MC-Val-Cit-PABC-MMAE drug-linker compound as described by Satomaa et al. [18]. Non-conjugated drug-linkers were removed by repeated additions of formulation buffer (i.e., 5% mannitol-0.1% Tween-PBS) and centrifugation through Amicon Ultracel 30 K centrifugal filter (Merck KGaA, Darmstadt, Germany). In order to characterize the compounds, 30 µg of each ADC was digested with FabRICATOR enzyme (Genovis, Lund, Sweden) and analyzed with MALDI-TOF mass spectrometry using sinapinic acid matrix. After this, the drug-to-antibody ratios were calculated based on the observed relative intensities of the light chain fragments (LC) at m/z 24926 (LC + MMAU), Fab heavy chain fragments (Fab-HC) at m/z 29854 (Fab-HC + 3 MMAU), and Fc heavy chain fragments (Fc) at m/z 25227, 25389, and 25551 for differentially galactosylated fragments G0F-Fc, G1F-Fc, and G2F-Fc, respectively [18]. Both ADCs were similarly analyzed and found to have a drug-to-antibody ratio of 8.

2.6. Cytotoxicity Assay

In vitro cytotoxicity of the free payloads and ADCs was assayed similarly as described before [9]. Briefly, a human ovarian cancer cell line SKOV-3 (ATCC, Manassas, VA, USA) was seeded in cell culture medium onto 96-well plates and incubated overnight in a cell incubator. Dilution series of free payloads and antibody-drug conjugates were applied to the cells in three parallel wells and the incubation was continued for 72 h. The viability of the cells was determined with PrestoBlue cell viability reagent (Life Technologies, Carlsbad, CA, USA) according to the manufacturer's instructions. The reagent was incubated with cells for 2–2.5 h and the absorbance was measured at 570 nm and 600 nm. IC_{50} values were determined using curve fitting by nonlinear regression as the concentration of the drug that causes 50% inhibition of cell viability compared to maximum inhibition.

3. Results and Discussion

3.1. Micellar Electrokinetic Chromatography

The hydrophobic character of the cytotoxic agents displayed in Figure 1 was assessed using MEKC. The technique is excellent for the separation of charged and neutral compounds, and particularly for assessing the hydrophobicity of analytes. In MEKC the fused silica separation capillary is filled with amphiphilic surfactants, which are able to coalesce to form micelles at concentrations exceeding the critical micelle concentration [19–22]. Generally, liposomes and micelles are accepted models for interpreting the interactions between molecules and lipid bilayers, and therefore studying the interaction of the payload molecules in this system creates a simplified model for studying the interactions of the liberated drugs with cellular membranes. This is undoubtedly an important factor for antibody-drug conjugates as a whole. Therefore, the results of the MEKC studies generate more information than theoretically calculated partitioning coefficients or distribution constants (i.e., logP or logD values), especially from a biological perspective.

To evaluate the hydrophobicity of the cytotoxic agents MMAE, 1, and 2 (see Figure 1), the retention factors (k) and, moreover, the distribution constants (K_D) of the compounds were determined using negatively charged, highly hydrophobic SDS micelles, and a mixture of SDS and less hydrophobic SC micelles as pseudostationary phase in MEKC. Since the distribution constant of a compound in electrokinetic chromatography illustrates the strength of interaction between the compound and the pseudostationary phase, the value directly reflects the hydrophobicity of the compound; the higher the interaction, the higher is the distribution constant, and consequently, the higher is the hydrophobicity of the compound. The distribution constants, and the logarithm of the distribution constants, of the compounds using 20 mM SDS and 20/40 mM SDS/SC dispersions were calculated using Equations (1)–(3) (in the experimental section) and the values are shown in Figure 2. The corresponding electropherograms are shown in Figure 3.

SDS is the most frequently used surfactant in MEKC, and therefore the hydrophobicity of the auristatins was initially assessed using solely SDS micelles. From the experimental data it was obvious that the attachment of carbohydrates containing hydrophilic functional groups to MMAE decreased the hydrophobicity of the parent molecule in the order of MMAE > 2 > 1. In more detail, the K_D value decreased by 13% when the N-terminal of MMAE was modified by reductive amination with 6-deoxy-6-azido-D-galactose to give compound 2, whereas the decrease was 36% when the benzylic hydroxyl group in the norephedrine residue was modified by glycosylation with glucuronic acid to give compound 1. The distribution constants, that is, the relative hydrophobicity of the cytotoxic agents, are logical. The modification of the N-terminal of MMAE, which gives rise to 2, adds a hydrophilic tail to the molecule containing four hydroxyl groups. Surprisingly, this modification does not significantly alter the hydrophobic character of MMAE. The attachment of a glucuronic acid to the benzylic hydroxyl group of MMAE resulting in 1 is associated with a greater increase in hydrophilic character, especially compared to 1. This is logical since glucuronic acid contains a carboxylic acid functionality in addition to the hydroxyl groups. A 36% decrease in the distribution constant is considerable, especially since the structural variations between 1 and 2 are relatively small.

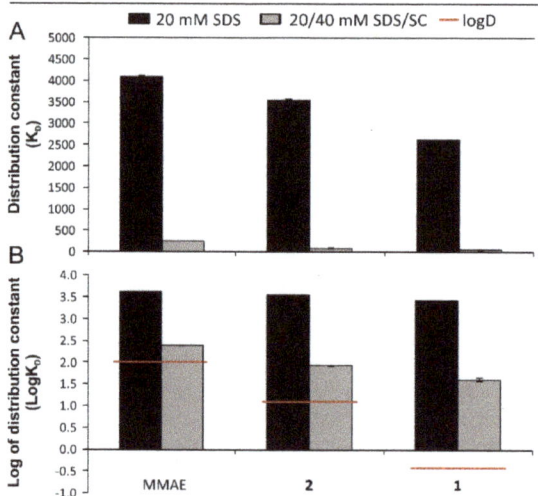

Figure 2. Distribution constants (displayed in 2A) and the logarithm of distribution constants (displayed in 2B) (including error bars) of MMAE, 1, and 2 by micellar electrokinetic chromatography (MEKC). Amounts of 20 mM sodium dodecyl sulfate (SDS) and 20 mM SDS mixed with 40 mM sodium cholate (SC) in sodium phosphate buffer (pH 7.4, I = 10 mM) were used as pseudostationary phases in MEKC. Thiourea was used as an EOF marker. Separation conditions were as follows: capillary length 30/38.5 cm; separation voltage 25 kV; capillary cassette temperature 25 °C; sample injection 10 s 10 mbar; UV detection at 200, 214, 238, and 254 nm. The inner and outer diameters of the capillary were 50 μm and 375 μm, respectively. The logD values represent the theoretically calculated distribution coefficients calculated by the ChemAxon software.

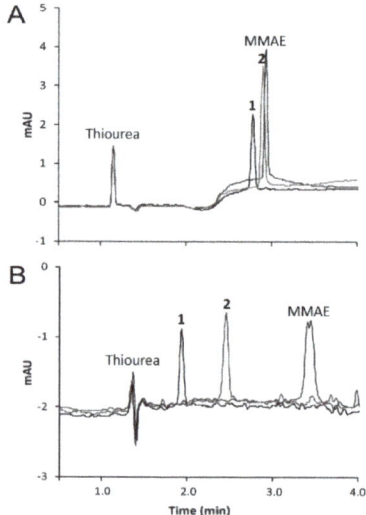

Figure 3. MEKC separation of the studied cytotoxic agents using (**A**) 20 mM SDS and (**B**) 20/40 mM of SDS/SC. The experimental conditions are similar to those in Figure 2.

The obtained results are consistent with theoretical octanol-water coefficients (logD) calculated by the ChemAxon software. The software estimates the logD values based on the structures of the compound as a function of pH, and the resulting logD values at pH 7.4 were 2.0, 1.1, and

−0.43 for MMAE, 2, and 1, respectively (shown as red lines in Figure 2). The logD values of the compounds follow the same order as the experimental logK_D values; however, the logD values were two to eight times lower and the differences between the logD values of the compounds were larger. The high experimentally determined logK_D values can be explained by the selection of SDS as the pseudostationary phase. SDS is a well-established anionic surfactant forming spherical micelles with a hydrophobic interior and a negatively charged hydrophilic exterior. These micelles are highly hydrophobic and, thus, the selectivity for neutral, highly hydrophobic, and fairly large compounds can be poor due to their nearly complete solubilization into the SDS micelles, resulting in co-migration of compounds [20,23–25]. In addition, SDS micelles are stronger hydrogen bond donors than 1-octanol and therefore they are expected to have strong interactions with hydrogen bond acceptor solutes like MMAE and its derivatives. Due to the poor selectivity of SDS micelles toward the used analytes, sodium cholate (SC), a biologically relevant anionic surfactant, was tested. SC is a common bile salt having a steroidal structure and it has been shown to be advantageous for separating hydrophobic compounds that cannot be separated by SDS [25–27].

Surprisingly, pure 40 mM SC micelles did not have any interaction with the cytotoxic agents (data not shown) and therefore different mixed micellar systems of SDS and SC were tested (20/10, 20/20, and 20/40 mM of SDS/SC). The addition of 10, 20, or 40 mM SC to the SDS solutions lowered the hydrophobicity of the SDS micelles in all cases, as witnessed by improved separations of the compounds with the optimal separation achieved by mixing 20 mM SDS with 40 mM SC. The compounds followed the same migration order as with pure SDS micelles (1 > 2 > MMAE), but the separation selectivity was much enhanced using a mixed SC and SDS system. With this pseudostationary phase, the hydrophobicity (K_D) decreased by 66% for compound 2 and by 84% for compound 1 (Figure 2B). Moreover, the logK_D values, obtained using the 20/40 mM SDS/SC mixture, were in a better correlation with the theoretical logD values than the values obtained using solely SDS as a pseudostationary phase (Figure 2B). One plausible explanation is that polar SC has a weaker solubilization power than SDS, and thus the SC/SDS mixture is a better model for biological systems [20].

3.2. Trastuzumab Conjugates and Cytotoxicity Assays

In our previous study, the cytotoxicities of MMAE (IC_{50} = 4 nM), 2 (IC_{50} = 12 nM), and an MMAE-glycolinker-cetuximab ADC (IC_{50} = 9 nm) were reported using HSC-2 head-and-neck squamous cell carcinoma cells [9]. It was rather surprising that the cytotoxicity displayed by MMAE was only marginally reduced when the N-terminal was modified with a glycolinker, since previous studies have established that an amide bond at the same site leads to the loss of cytotoxic activity [28]. While stable linkers have been found to improve the tolerability and antitumor activity in other state-of-the art ADCs (e.g., the anti-HER2 ADC trastuzumab emtansine) [29], the glycolinker strategy failed to yield a superior ADC and the approach was not considered competitive enough.

Therefore, we decided to use a cleavable linker in this study. While a number of cleavable linkers exist (e.g., pH-sensitive linkers, enzyme targeting linkers) [25,26], we opted to use the cathepsin B cleavable linker valine-citrulline-*p*-aminocarbamate (Val-Cit-PABC) which is currently used in brentuximab vedotin [30]. Furthermore, we decided to evaluate the effects of the hydrophilic glucuronic acid residue at the hydroxyl group in the norephedrine residue of MMAE (the MMAE-glycolinker-substrate was thus not re-evaluated). In addition, it should be noted that the proteolysis of the Val-Cit-PABC-linker releases the warhead molecule as such instead of a conjugate thereof [31]. Therefore, it was important to start by evaluating the cytotoxicity of MMAE and MMAU (see Figure 1).

SKOV-3 ovarian carcinoma cells expressing the HER2 receptor protein were used, and the results are summarized in Figure 4. When evaluating the cytotoxicity using this cell line, MMAU was over three orders of magnitude less cytotoxic than MMAE when applied to the cell culture medium. It is therefore clear that an increase in the hydrophilic character of the MMAE derivatives is accompanied by a decrease in the cytotoxic activity displayed. Based on these results alone, increasing the hydrophilicity and a steric bulk of the norephedrine-residue leads to a substantial decrease in the cytotoxicity of the parent molecule in contrast to the previous hydrophilic modification of the N-terminal. This may be related to the cellular uptake mechanism, which favors hydrophobic molecules, or to the binding mode by which the auristatins act at the tubulin receptor [32,33]. The first mechanism seems more plausible since we have established that MMAU displays an effective bystander effect when conjugated to an internalizing antibody such as an ADC [18]. This further indicates that the glucuronide is hydrolyzed in the intracellular milieu thus liberating the actual drug, namely, MMAE.

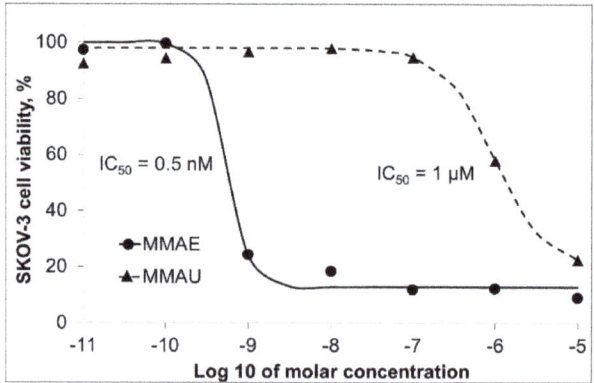

Figure 4. Cytotoxicity assays reveal that the hydrophobic payload MMAE is 2000-fold more cytotoxic against SKOV-3 ovarian cancer cells than MMAU.

In a traditional study on cytotoxic agents, these results would be conclusive; however, ADCs function by a different mechanism, that is, by the internalization of the entire ADC, followed by the release of the warhead molecule. Therefore, studying the properties of the cytotoxic agents alone does not provide the complete picture. As a result, we decided to conduct a preliminary study with ADCs featuring MMAE and MMAU. Since we used HER2-expressing ovarian cancer cells, we chose to use trastuzumab as the antibody. Trastuzumab was a logical choice since it is currently applied in both its naked form and as an ADC (trastuzumab emtansine) [34] in the treatment of HER2-positive breast cancer patients.

The strategy employed in the construction of the ADCs is displayed in Scheme 1 and described in detail elsewhere [18]. This resulted in compounds 3 and 4 being, respectively, the trastuzumab-MC-Val-Cit-PABC-MMAE derivative and the corresponding MMAU-trastuzumab conjugate. The preliminary cytotoxicity studies were conducted with an HER2-positive SKOV-3 cancer cell line (results summarized in Figure 5). Surprisingly, both compounds 3 and 4 displayed similar cytotoxicities with IC_{50} values in the pM-range (50–80 pM). The results are interesting, especially due to the large deviation in the cytotoxic activity of the free warheads. Taken together, the results indicate that the trastuzumab-MC-Val-Cit-PABC-MMAU derivative would be equally toxic as the corresponding MMAE-trastuzumab conjugate, however with a potentially decreased amount of off-site toxic side effects due to the reduced toxicity of the free drug.

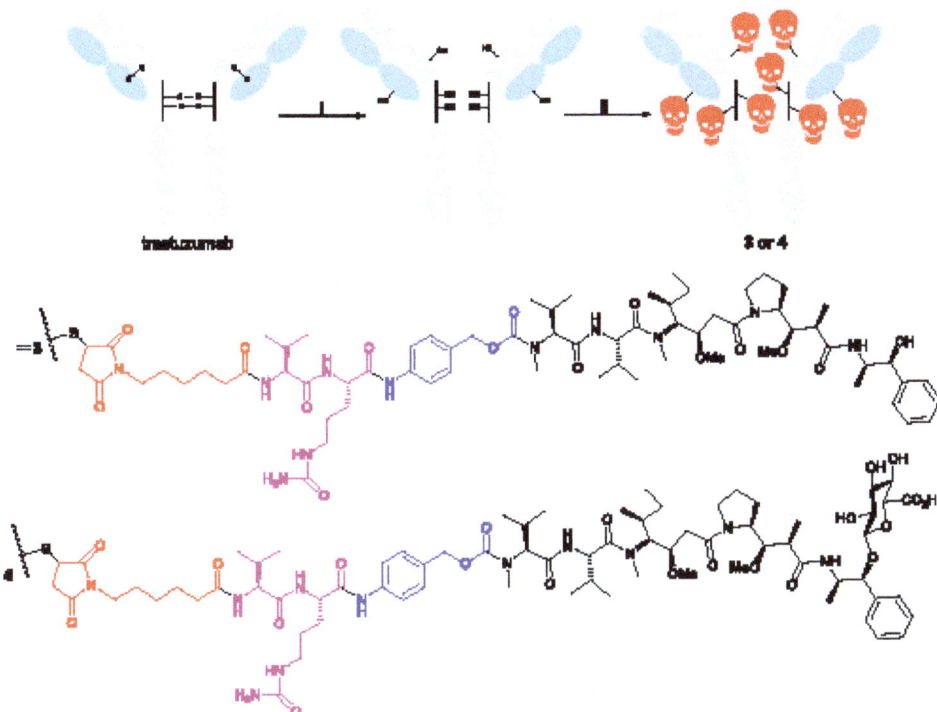

Scheme 1. Schematic view of the construction of trastuzumab-MC-Val-Cit-PABC-auristatin ADCs. (i) TCEP reduction of interchain disulfides; (ii) thiol-maleimide conjugation of payload molecules. The warhead–linker conjugates are displayed below the conjugation route with different parts highlighted; the cytotoxic warhead molecule (black), the self-immolative PABC-spacer (blue), the cathepsin B cleavable Val-Cit-dipeptide (purple), and the MC conjugation site (red).

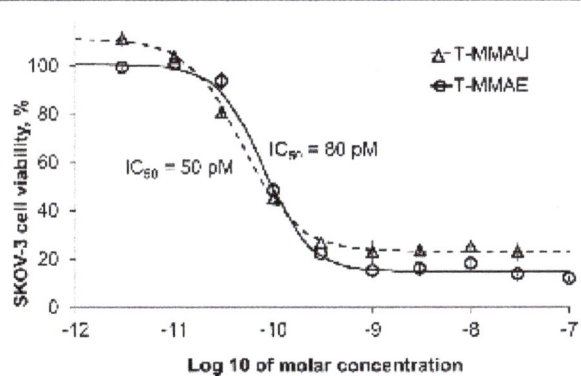

Figure 5. Trastuzumab antibody (T)-payload conjugates 3 and 4 have nearly comparable cytotoxicity against the HER2-expressing SKOV-3 cells, whereas they are between 10-fold (against MMAE) and 20,000-fold (against MMAU) more cytotoxic than the free payloads. In the assays, different concentrations of ADCs were applied to the cell culture medium and the cell viability was evaluated after 72 h incubation. The graph shows the average values of three parallel experiments. Error bars show the standard deviation.

These results prove that the construction of competitive ADCs can be achieved by the use of cytotoxic warheads that would normally be disregarded based on cytotoxicity assays alone. On a more fundamental level, the question that arises is why ADCs are not regularly screened for the delivery of hydrophilic agents across the cellular membrane, that is, molecules that would normally not pass the membrane. This is an area which requires further research and as our preliminary study on the cytotoxicity of the trastuzumab-auristatin ADCs shows, there is a significant potential for the development of novel ADCs with hydrophilic payload molecules. While we have focused on the use of carbohydrates as a means of increasing the hydrophilicity, other strategies have also been reported (e.g., the use of charged phosphate groups) [35]. It is clear that there is a need for more systematic studies with larger molecule libraries featuring hydrophilic and hydrophobic warhead molecules in order to provide detailed insights on the subject. Nonetheless, the benefits of carbohydrates are related to their specific lysosomal cleavage pathways in human cells, which provide an elegant way of adding hydrophilicity to the ADCs during the systemic distribution and localization in target tissues. Simultaneously, and using this pro-drug strategy, the fully active hydrophobic cytotoxic agent is generated after ADC internalization in target cells. In a recent follow-up study, we have shown that this concept is viable, since MMAU-ADCs show excellent cytotoxic activity against cancer cells both in vitro and in vivo, where complete eradication of tumors in xenografted mice was achieved [18].

4. Conclusions

MEKC was used for determining the relative hydrophobicities of modified auristatins. Optimal separation was achieved using a mixed micellar system of 20 mM SDS and 40 mM SC as the pseudostationary phase. The hydrophobicity of the auristatins decreased in the order of MMAE > compound 2 > compound 1, and the same pattern was identified in the cytotoxic assays where the cytotoxicity decreased in the same order.

Hydrophilic modification (attachment of a glucuronic acid residue) at the norephedrine residue of MMAE (compound 1; MMAU) was accompanied by a significant decrease in the cytotoxicity. The addition of a hydrophilic glycolinker at the N-terminal, on the other hand, had only a slight effect on the cytotoxicity (compound 2), which is an interesting observation since an amide bond at this position is known to render the auristatins close to non-toxic.

The preliminary cytotoxic evaluation of the trastuzumab-MC-Val-Cit-PABC-auristatin conjugates 3 and 4 revealed IC_{50}-values in the pM range. These values are comparable to other leading state-of-the-art ADCs. Furthermore, the MMAU-ADCs have certain advantages over the MMAE-ADCs, for example, prematurely cleaved linkers will be accompanied by less off-site harmful side effects due to the lower cytotoxicity of MMAU when compared to MMAE. Altogether, our results on the use of hydrophilic payload conjugates prove that novel opportunities exist for the future design of ADCs with previously neglected hydrophilic molecules.

Author Contributions: Conceptualization, F.S.E., S.-K.R., S.K.W.; Investigation, F.S.E., S.-K.R., M.R., V.P., A.V., J.S., J.H.; Resources, F.S.E., T.S., S.K.W.; Writing—Original Draft Preparation, F.S.E., S.-K.R., T.S., S.K.W.; Writing—Review and Editing, F.S.E., S.-K.R., S.K.W., T.S.; Visualization, F.S.E., S.-K.R., S.K.W.; Supervision, S.K.W.; Project Administration, F.S.E., S.K.W.; Funding Acquisition, F.S.E., S.K.W.

Funding: This research was funded by the Magnus Ehrnrooth Foundation (S.K.W.), the Ruth and Nils-Erik Stenbäck Foundation (F.S.E), The Finnish Society of Sciences and Letters (S.K.W.), and the Academy of Finland (S.K.W; project number 266342).

Acknowledgments: Jesper Långbacka is acknowledged for assistance with the CMC measurements.

Conflicts of Interest: The authors declare no conflict of interest. The founding sponsors had no role in the design of the study; in the collection, analyses, or interpretation of data; in the writing of the manuscript; and in the decision to publish the results.

References

1. Schrama, D.; Reisfeld, R.A.; Becker, J.C. Antibody targeted drugs as cancer therapeutics. *Nat. Rev. Drug Discov.* **2006**, *5*, 147–159. [CrossRef] [PubMed]
2. Alley, S.C.; Okeley, N.M.; Senter, P.D. Antibody–drug conjugates: targeted drug delivery for cancer. *Curr. Opin. Chem. Biol.* **2010**, *14*, 529–537. [CrossRef] [PubMed]
3. Younes, A.; Bartlett, N.L.; Leonard, J.P.; Kennedy, D.A.; Lynch, C.M.; Sievers, E.L.; Forero-Torres, A. Brentuximab vedotin (SGN-35) for relapsed CD30-positive lymphomas. *N. Eng. J. Med.* **2010**, *363*, 1812–1821. [CrossRef] [PubMed]
4. Verma, S.; Miles, D.; Gianni, L.; Krop, I.E.; Welslau, M.; Baselga, J.; Pegram, M.; Oh, D.-Y.; Diéras, V.; Guardino, E. Trastuzumab emtansine for HER2-positive advanced breast cancer. *N. Eng. J. Med.* **2012**, *367*, 1783–1791. [CrossRef] [PubMed]
5. Rowe, J.M.; Löwenberg, B. Gemtuzumab ozogamicin in acute myeloid leukemia: a remarkable saga about an active drug. *Blood* **2013**, *121*, 4838–4841. [CrossRef] [PubMed]
6. Rytting, M.; Triche, L.; Thomas, D.; O'brien, S.; Kantarjian, H. Initial experience with CMC-544 (inotuzumab ozogamicin) in pediatric patients with relapsed B-cell acute lymphoblastic leukemia. *Pediatr. Blood Cancer* **2014**, *61*, 369–372. [CrossRef] [PubMed]
7. Lyon, R.P.; Bovee, T.D.; Doronina, S.O.; Burke, P.J.; Hunter, J.H.; Neff-LaFord, H.D.; Jonas, M.; Anderson, M.E.; Setter, J.R.; Senter, P.D. Reducing hydrophobicity of homogeneous antibody-drug conjugates improves pharmacokinetics and therapeutic index. *Nat. Biotechnol.* **2015**, *33*, 733–735. [CrossRef] [PubMed]
8. Zhao, R.Y.; Wilhelm, S.D.; Audette, C.; Jones, G.; Leece, B.A.; Lazar, A.C.; Goldmacher, V.S.; Singh, R.; Kovtun, Y.; Widdison, W.C. Synthesis and evaluation of hydrophilic linkers for antibody–maytansinoid conjugates. *J. Med. Chem.* **2011**, *54*, 3606–3623. [CrossRef] [PubMed]
9. Ekholm, F.S.; Pynnönen, H.; Vilkman, A.; Pitkänen, V.; Helin, J.; Saarinen, J.; Satomaa, T. Introducing glycolinkers for the functionalization of cytotoxic drugs and applications in antibody–drug conjugation chemistry. *ChemMedChem* **2016**, *11*, 2501–2505. [CrossRef]
10. Johansson, M.P.; Maaheimo, H.; Ekholm, F.S. New insight on the structural features of the cytotoxic auristatins MMAE and MMAF revealed by combined NMR spectroscopy and quantum chemical modelling. *Sci. Rep.* **2017**, *7*, 15920. [CrossRef] [PubMed]
11. Rostami, S.; Qazi, I.; Sikorski, R. The clinical landscape of antibody-drug conjugates. *ADC Rev.* **2014**. [CrossRef]
12. Barok, M.; Joensuu, H.; Isola, J. Trastuzumab emtansine: mechanisms of action and drug resistance. *Breast Cancer Res.* **2014**, *16*, 209. [CrossRef] [PubMed]
13. Hamblett, K.J.; Senter, P.D.; Chace, D.F.; Sun, M.M.; Lenox, J.; Cerveny, C.G.; Kissler, K.M.; Bernhardt, S.X.; Kopcha, A.K.; Zabinski, R.F. Effects of drug loading on the antitumor activity of a monoclonal antibody drug conjugate. *Clin. Cancer Res.* **2004**, *10*, 7063–7070. [CrossRef] [PubMed]
14. Bushey, M.M.; Jorgenson, J.W. Separation of dansylated methylamine and dansylated methyl-d3-amine by micellar electrokinetic capillary chromatography with methanol-modified mobile phase. *Anal. Chem.* **1989**, *61*, 491–493. [CrossRef]
15. Laine, J.; Lokajová, J.; Parshintsev, J.; Holopainen, J.M.; Wiedmer, S.K. Interaction of a commercial lipid dispersion and local anesthetics in human plasma: Implications for drug trapping by "lipid-sinks". *Anal. Bioanal. Chem.* **2010**, *396*, 2599–2607. [CrossRef] [PubMed]
16. Ahlstrom, D.M.; Hoyos, Y.M.; Arslan, H.; Akbay, C. Binary mixed micelles of chiral sodium undecenyl leucinate and achiral sodium undecenyl sulfate: I. Characterization and application as pseudostationary phases in micellar electrokinetic chromatography. *J. Chromatogr. A* **2010**, *1217*, 375–385. [CrossRef] [PubMed]
17. González-Gaitano, G.; Compostizo, A.; Sánchez-Martín, L.; Tardajos, G. Speed of sound, density, and molecular modeling studies on the inclusion complex between sodium cholate and β-cyclodextrin. *Langmuir* **1997**, *13*, 2235–2241. [CrossRef]
18. Satomaa, T.; Pynnönen, H.; Vilkman, A.; Kotiranta, T.; Pitkänen, V.; Heiskanen, A.; Herpers, B.; Price, L.S.; Helin, J.; Saarinen, J. Hydrophilic Auristatin Glycoside Payload Enables Improved Antibody-Drug Conjugate Efficacy and Biocompatibility. *Antibodies* **2018**, *7*, 15. [CrossRef]
19. Terabe, S.; Otsuka, K.; Ichikawa, K.; Tsuchiya, A.; Ando, T. Electrokinetic separations with micellar solutions and open-tubular capillaries. *Anal. Chem.* **1984**, *56*, 111–113. [CrossRef]

20. Nishi, H.; Terabe, S. Micellar electrokinetic chromatography perspectives in drug analysis. *J. Chromatogr. A* **1996**, *735*, 3–27.
21. Silva, M. Micellar electrokinetic chromatography: A review of methodological and instrumental innovations focusing on practical aspects. *Electrophoresis* **2013**, *34*, 141–158. [CrossRef] [PubMed]
22. Deeb, S.E.; Dawwas, H.A.; Gust, R. Recent methodological and instrumental development in MEKC. *Electrophoresis* **2013**, *34*, 1295–1303. [CrossRef] [PubMed]
23. Ji, A.J.; Nunez, M.F.; Machacek, D.; Ferguson, J.E.; Iossi, M.F.; Kao, P.C.; Landers, J.P. Separation of urinary estrogens by micellar electrokinetic chromatography. *J. Chromatogr. B* **1995**, *669*, 15–26. [CrossRef]
24. Jumppanen, J.H.; Wiedmer, S.K.; Siren, H.; Riekkola, M.L.; Haario, H. Optimized Separation of 7 Corticosteroids by Micellar Electrokinetic Chromatography. *Electrophoresis* **1994**, *15*, 1267–1272. [CrossRef] [PubMed]
25. Wiedmer, S.K.; Jumppanen, J.H.; Haario, H.; Riekkola, M.L. Optimization of selectivity and resolution in micellar electrokinetic capillary chromatography with a mixed micellar system of sodium dodecyl sulfate and sodium cholate. *Electrophoresis* **1996**, *17*, 1931–1937. [CrossRef] [PubMed]
26. Cole, R.O.; Sepaniak, M.J.; Hinze, W.L.; Gorse, J.; Oldiges, K. Bile salt surfactants in micellar electrokinetic capillary chromatography: application to hydrophobic molecule separations. *J. Chromatogr. A* **1991**, *557*, 113–123. [CrossRef]
27. Yang, S.; Bumgarner, J.G.; Kruk, L.F.; Khaledi, M.G. Quantitative structure-activity relationships studies with micellar electrokinetic chromatography influence of surfactant type and mixed micelles on estimation of hydrophobicity and bioavailability. *J. Chromatogr. A* **1996**, *721*, 323–335. [CrossRef]
28. Doronina, S.O.; Mendelsohn, B.A.; Bovee, T.D.; Cerveny, C.G.; Alley, S.C.; Meyer, D.L.; Oflazoglu, E.; Toki, B.E.; Sanderson, R.J.; Zabinski, R.F. Enhanced activity of monomethylauristatin F through monoclonal antibody delivery: effects of linker technology on efficacy and toxicity. *Bioconjug. Chem.* **2006**, *17*, 114–124. [CrossRef]
29. Lambert, J.M.; Chari, R.V. Ado-trastuzumab Emtansine (T-DM1): an antibody–drug conjugate (ADC) for HER2-positive breast cancer. *J. Med. Chem.* **2014**, *57*, 6949–6964. [CrossRef]
30. Alley, S.C.; Benjamin, D.R.; Jeffrey, S.C.; Okeley, N.M.; Meyer, D.L.; Sanderson, R.J.; Senter, P.D. Contribution of linker stability to the activities of anticancer immunoconjugates. *Bioconjug. Chem.* **2008**, *19*, 759–765. [CrossRef]
31. Senter, P.D.; Sievers, E.L. The discovery and development of brentuximab vedotin for use in relapsed Hodgkin lymphoma and systemic anaplastic large cell lymphoma. *Nat. Biotechnol.* **2012**, *30*, 631–637. [CrossRef] [PubMed]
32. Waight, A.B.; Bargsten, K.; Doronina, S.; Steinmetz, M.O.; Sussman, D.; Prota, A.E. Structural basis of microtubule destabilization by potent auristatin anti-mitotics. *PLoS ONE* **2016**, *11*, e0160890. [CrossRef] [PubMed]
33. Wang, Y.; Benz, F.W.; Wu, Y.; Wang, Q.; Chen, Y.; Chen, X.; Li, H.; Zhang, Y.; Zhang, R.; Yang, J. Structural Insights into the Pharmacophore of Vinca Domain Inhibitors of Microtubules. *Mol. Pharmacol.* **2016**, *89*, 233–242. [CrossRef] [PubMed]
34. Phillips, G.D.L.; Li, G.; Dugger, D.L.; Crocker, L.M.; Parsons, K.L.; Mai, E.; Blättler, W.A.; Lambert, J.M.; Chari, R.V.; Lutz, R.J. Targeting HER2-positive breast cancer with trastuzumab-DM1, an antibody–cytotoxic drug conjugate. *Cancer Res.* **2008**, *68*, 9280–9290. [CrossRef] [PubMed]
35. Zhao, R.Y.; Erickson, H.K.; Leece, B.A.; Reid, E.E.; Goldmacher, V.S.; Lambert, J.M.; Chari, R.V. Synthesis and biological evaluation of antibody conjugates of phosphate prodrugs of cytotoxic DNA alkylators for the targeted treatment of cancer. *J. Med. Chem.* **2012**, *55*, 766–782. [CrossRef] [PubMed]

 © 2018 by the authors. Licensee MDPI, Basel, Switzerland. This article is an open access article distributed under the terms and conditions of the Creative Commons Attribution (CC BY) license (http://creativecommons.org/licenses/by/4.0/).

Article

A Low-Cost Approach Using Diatomaceous Earth Biosorbent as Alternative SPME Coating for the Determination of PAHs in Water Samples by GC-MS

Naysla Paulo Reinert [1], Camila M. S. Vieira [1], Cristian Berto da Silveira [2], Dilma Budziak [3] and Eduardo Carasek [1,*]

[1] Departamento de Química, Universidade Federal de Santa Catarina, Florianópolis 88040-900, SC, Brazil; nayslareinert@gmail.com (N.P.R.); camilamaiara.vieira@gmail.com (C.M.S.V.)
[2] Departamento de Engenharia de Pesca e Ciências Biológicas, Universidade do Estado de Santa Catarina, Laguna, Santa Catarina 88790-000, Brazil; cristian.silveira@udesc.br
[3] Departamento de Ciências Naturais e Sociais, Universidade Federal de Santa Catarina, Curitibanos 89520-000, SC, Brazil; dilmabudziak@yahoo.com.br
* Correspondence: eduardo.carasek@ufsc.br

Received: 10 October 2018; Accepted: 12 November 2018; Published: 20 November 2018

Abstract: In this study, the use of recycled diatomaceous earth as the extraction phase in solid phase microextraction (SPME) technique for the determination of polycyclic aromatic hydrocarbons (PAHs) in river water samples, with separation/detection performed by gas chromatography-mass spectrometry (GC-MS), is proposed. The optimized extraction conditions are extraction time 70 min at 80 °C with no addition of salt. The limits of quantification were close to 0.5 µg L^{-1} with RSD values lower than 25% (n = 3). The linear working range was 0.5 µg L^{-1} to 25 µg L^{-1} for all analytes. The method was applied to samples collected from the Itajaí River (Santa Catarina, Brazil) and the RSD values for repeatability and reproducibility were lower than 15% and 17%, respectively. The efficiency of the recycled diatomaceous earth fiber was compared with that of commercial fibers and good results were obtained, confirming that this is a promising option to use as the extraction phase in SPME.

Keywords: recycled diatomaceous earth; solid phase microextraction; polycyclic aromatic hydrocarbons; gas chromatography-mass spectrometry

1. Introduction

Water is an extremely valuable natural resource as it is responsible for maintaining biological, geological and chemical cycles [1,2]. Environmental problems caused by anthropogenic activities are continually increasing and gaining attention worldwide [1]. With population growth and increased industrial activities, ever greater amounts of petroleum-based fossil fuels are being consumed [3]. These fuels contain a class of compounds known as polycyclic aromatic hydrocarbons (PAHs).

PAHs are a group of organic compounds composed of multiple aromatic rings [4]. The formation of these molecules is associated with the incomplete combustion of natural organic materials, for instance, due to volcanoes or the incomplete burning of wood in forest fires, and from anthropogenic sources including industrial processes (e.g., refineries), vehicular emissions [5], cane burning [6], and others [7]. According to the International Agency for Research on Cancer (IARC) and the US EPA (the United States Environmental Protection Agency) PAHs are recognized as persistent environmental pollutants with carcinogenic and mutagenic capacity in humans [7,8]. Based on these issues, measures have been taken by governments around the world to monitor the concentrations of compounds that may be harmful to human health, with different standards and regulations being established often aimed at ensuring the quality of drinking water [7]. In Brazil, the Ministry of Health regulates waters for human consumption using benzo[a]pyrene as a marker with maximum permitted values of 0.7 µg L^{-1}.

The determination of these pollutants generally requires a sample preparation procedure to remove matrix interferents, concentrate the analyte and make the extract compatible with the analytical instrumentation. One of the most commonly used sample preparation techniques is solid-phase microextraction (SPME) [9,10].

SPME was proposed by Pawliszyn et al. in 1990 to overcome the drawbacks of traditional sample preparation techniques such as liquid-liquid extraction and solid phase extraction [9,10]. The principle of the technique is the distribution of the analytes between the sample matrix and the sorbent (fiber), combining sampling, isolation and enrichment in a single step [11,12]. SPME fibers are composed of a fused silica or metallic support coated with an extractive phase, for instance, polymethylsiloxane (PDMS), polyacrylate (PA), or other commercially available sorbent [13–16].

In the search for new sorbent materials for SPME, biosorbents have gained prominence in miniaturized techniques because they provide greener, less expensive, renewable, and biodegradable extractive phases. Many of these biosorbents can be found in the environment and consist of macromolecules with different functional groups that can interact with different types of analytes. Our research group has previously used natural sorbents for the determination of organic contaminants using SPME [17,18]. Diatomaceous earth is of particular interest as a new biosorbent since it is discarded in large scale as a waste from breweries, where it is used for the clarification and filtration of organic materials and beers [19].

Diatomaceous earth is obtained from sedimentary rocks, originating from fossilized algae belonging to the class *Bacillariophyta* (diatoms). It is an amorphous mineral, comprised mainly of silica dioxide, of light weight and low molar mass, and its coloration can vary from white to gray. Structurally, diatoms have a hollow cylindrical form of low density and high surface area [20].

In this study the use of diatomaceous earth as an (bio) extractive phase in SPME is explored for the determination of PAHs in river water samples with quantification by gas chromatography coupled to mass spectrometry (GC-MS). The biosorbent was easily adhered onto a NiTi (nitinol) rod using a quick and inexpensive procedure.

2. Materials and Methods

2.1. Reagents and Materials

Analytical standards of PAHs in a mixture containing acenaphthylene, fluorene, phenanthrene, anthracene, pyrene, benzo[a]anthracene, chrysene, benzo[b]fluoranthene, benzo[k]fluoranthene, and benzo[a]pyrene (Bellefonte, PA, USA) were used to prepare stock solutions of 1 mg L^{-1} in acetonitrile purchased from J.T. Baker (Mallinckrodt, NJ, USA). The ionic strength was studied using sodium chloride obtained from Synth (São Paulo, SP, Brazil). The ultrapure water used in the experiments was purified in an ultrapure Mega purity system (Billerica, MA, USA). The fiber was prepared using diatomaceous earth with size less than 200 mesh, nitinol rods (2 cm length and 0.128 mm diameter), epoxy glue acquired from Brascola (São Paulo, SP, Brazil), and a heating block from Dist (Florianópolis, SC, Brazil). SPME extractions were carried out in vials of 40 mL obtained from Supelco (Bellefonte, PA, USA) aided by a thermostatic bath (Lab Companion RW 0525G, Geumcheon-gu, Seoul, Korea) and magnetic stirrers from Dist (Florianópolis, SC, Brazil). Commercial fibers (DVB/Car/PDMS, 50/30 µm; PDMS 100 µm and PDMS/DVB, 65 µm; Supelco, Bellafonte, PA, USA) were used to compare the analyte extraction efficiencies.

2.2. Instrumental and Chromatographic Conditions

An Agilent 7820A gas chromatograph with flame ionization detector (FID) equipped with a split/splitless injector and an Agilent DB-5 capillary column (30 m × 0.25 mm × 0.25 µm; Santa Clara,

CA, USA) was used to optimize the method as well as to compare it with commercial fibers. On the other hand, a Shimadzu GC-MS QP2010 Plus equipped with a split/splitless injector (Kyoto, Japan) containing a Zebron ZB-5MS capillary column (30 m × 0.25 mm × 0.25 µm; Torrance, CA, USA) was used to obtain the analytical parameters of merit. The GC-MS and GC/FID was operated at the same conditions for injection and the columns temperature programs. The injection was performed in splitless mode at 260 °C for 15 min. The column temperature program consisted of maintaining the oven at 80 °C for 1 min and then increasing it 6 °C min^{-1} to 300 °C which was maintained for 10 min. The transfer line temperature, the ion source temperature and the electron impact ionization (EI) mode of the GC-MS were set at 280, 250 °C, and 70 eV, respectively. Helium was used as the carrier gas at a flow rate of 1.0 mL min^{-1}. The PAHs were determined in selected ion monitoring (SIM) mode and the mass/charge (m/z) ratios employed are shown in Table 1. The m/z values in bold were used for the quantitative determination of the analytes.

Table 1. The m/z values used for the determination of PAHs by GC-MS (values in bold were used for the quantification of the analytes).

Analytes	m/z
acenaphthylene	**152**, 153, 151
fluorene	**166**, 165, 167
phenanthrene	**178**, 176, 179
anthracene	**178**, 179, 176
pyrene	**202**, 203, 200
benzo[a]anthracene	**228**, 226, 229
chrysene	**228**, 226, 229
benzo[b]fluoranthene	**252**, 250, 126
benzo[k]fluoranthene	**252**, 250, 126
benzo[a]pyrene	**252**, 250, 126

2.3. Preparation of Diatomaceous Earth Fibers

The diatomaceous earth dust came from the disposal reservoir of a brewery, where this material is used for the filtration and clarification of beer (Santa Catarina, Brazil). Due to its high porosity, the material presents a high degree of saturation with organic matter from the treatment of beer. Thus, a heat treatment is required [20], not only to eliminate the residues originated from the beer filtration but to ensure that all of the organic matter adhered to the material is removed. The diatomaceous earth, after the thermal treatment, was sieved to obtain homogeneous particle size (<200 mesh). The diatomaceous earth was adhered on a 1 cm nitinol wire using epoxy glue. Then, the new fiber was inserted into the heating block at 180 °C for 90 min, resulting in a final phase thickness of approximately 40 µm. The fiber was then conditioned at 240 °C for 90 min in a GC injection port. The fiber lifetime was verified during the study by comparing the responses of the chromatographic areas of the analytes to the optimum extraction condition at a concentration of 5 µg L^{-1}. Fibers were used while the extraction efficiency did not present a reduction greater than 10%.

2.4. Optimization of SPME Procedure

The optimization of the extraction conditions for the diatomaceous earth fiber was performed by multivariate procedures. A central composite design involving 11 experiments with triplicate at the central point was carried out. In the optimization strategy the extraction temperature ranged from 30 to 80 °C and the extraction time from 30 to 117 min. The sodium chloride concentration (0–20% m/v) was also evaluated, but in the univariate form. The extraction procedure consisted of immersing the SPME fiber directly in 25 mL of water sample spiked with 100 µg L^{-1} of each PAH contained in a 40 mL vial and kept under constant magnetic stirring at 1000 rpm. After the extraction, the fiber was immediately inserted into the GC injection port at 240 °C for 15 min for the thermal desorption of the analytes. The analysis was carried out by GC-FID in splitless mode. To obtain the response surface, the geometric mean of the areas of the chromatographic peaks obtained in each extraction using Statistica 8.0 software (Statsoft, USA) was used.

2.5. Comparison of the Extraction Efficiencies Using Diatomaceous Earth and Commercial Fibers

After the optimization of the analytical procedure, the diatomaceous earth was compared to commercial fibers (PDMS and PDMS/DVB) in terms of their efficiency in the extraction of the PAHs studied. The same procedure described at Section 2.4 was carried out but the ultrapure water was spiked with the analytes at a concentration of 5 µg L^{-1}. The extractions were performed using one of the fibers at 80 °C for 70 min. The chromatographic analysis was performed by GC-MS.

2.6. Analytical Figures of Merit of the Method Developed

River water spiked with five concentrations of each analyte ranging from 0.5 to 25.0 µg L^{-1} was prepared to build calibration curves which were used to calculate the linear coefficient of determination (R^2). The lowest concentration on the analytical curve for each analyte which enabled measures with acceptable precision (RSD < 20%) was adopted as limits of quantification (LOQs). The limits of detection (LODs) were obtained dividing the LOQ by 3.3. The precision and the accuracy of the method were evaluated by performing extractions using real water samples spiked with the analytes at 0.5 µg L^{-1}. Precision was calculated as the relative standard deviation (RSD) obtained from spiked river water and accuracy was verified through the relative recovery of the analytes.

3. Results and Discussion

3.1. Characterization of the Diatomaceous Fiber

The diatomaceous earth samples used for the production of SPME fibers belong to the class *Bacillariophyceae centricae* and their color may vary from white to gray. The material consists mainly of silica, SiO_2 (87–91%), alumina and ferric oxide [21].

Scanning electron microscopy (SEM) was carried out to characterize the surface morphology of the recycled diatomaceous earth. The images obtained at magnifications of 2000 and 4000× for the surface evaluation are shown in Figure 1 (A and B, respectively). An image of a cross-section of the proposed fiber was obtained at a magnification of 100× (Figure 1C). According to the SEM results, the morphology of the material shows a high porosity which facilitated the physical processes involving the sorption of the analytes.

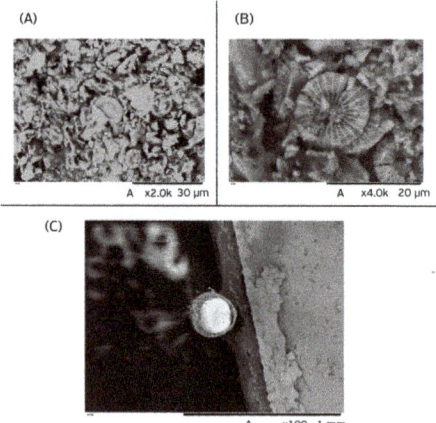

Figure 1. SEM micrographs obtained with the biosorbent fiber at magnifications of (**A**) 2000× and (**B**) 4000×, and a cross-section of the proposed fiber (**C**) at a magnification of 100×.

FTIR spectroscopy was carried out to identify the functional groups in the sorbent. The FTIR spectrum obtained from the material previously conditioned at 240 °C is illustrated in Figure 2. A broad peak at ~3400 cm^{-1} corresponds to the O–H bonds of silanol groups. Two intense peaks between ~1200 and 1080 cm^{-1} were assigned to the asymmetric stretching of the Si–O–Si siloxane groups and one at ~790 cm^{-1} is related to the Si–O–Si vibrations attributed to mesoporous silicas. At ~475 cm^{-1}, a peak related to O–Si–O vibration was present. Lastly, the peak at ~1600 cm^{-1} refers to the angular deformation of the adsorbed water molecules.

Thermogravimetric analysis was conducted to identify if there was any organic material present in the sample and since no mass loss was observed the material can be characterized as thermally stable (data not shown). This result was already expected, since the sorbent comes from inorganic material.

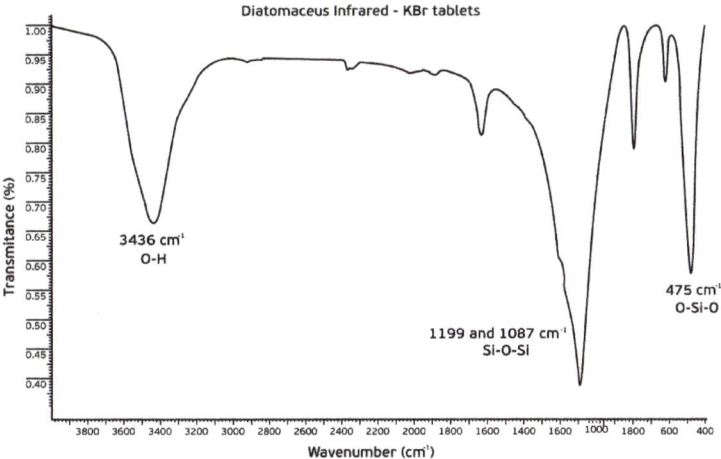

Figure 2. FTIR spectrum of the biosorbent, previously conditioned at 240 °C.

3.2. Optimization of DI-SPME Extraction Procedure

The extraction conditions that can influence the SPME efficiency were optimized using the diatomaceous earth fiber. The response used to feed the software Statistica 8.0 was the geometric means of the chromatographic peak areas of the analytes. The response surfaces obtained for the biosorbent fiber are shown in Figure 3.

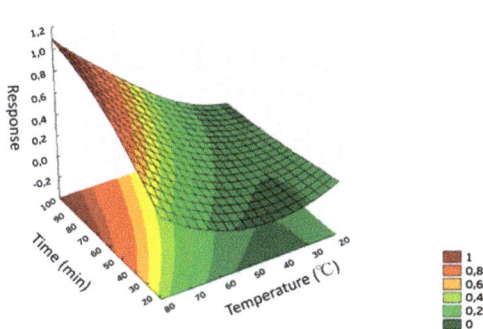

Figure 3. Response surface obtained for the optimization of DI-SPME procedure using biosorbent fiber (diatomaceous earth).

The optimum extraction conditions selected for the proposed fiber were reached using an extraction time of 70 min at 80 °C. The addition of salt was also studied as it is known to lead to the salting-out effect. However, the use of small amounts of salt caused fiber damage and so no salt was added in the extractions.

3.3. Comparison between the Extraction Efficiencies of the Biosorbent and Commercial Coatings

A comparison between the extraction efficiencies using the proposed fiber and commercial fibers (PDMS/DVB and PDMS) was performed. The conditions for the extractions using commercial fibers were optimized (data not shown) as extraction time of 70 min at 80 °C. These values are much closed to those mentioned in the literature (extractions of 60 min at 70 °C) [22–24]. Figure 4 shows this comparison through bar graph using normalized peak area and considering the film thickness of each fiber. The normalization of peak areas for each analyte was made using the highest chromatographic peak areas as 100% for each analyte.

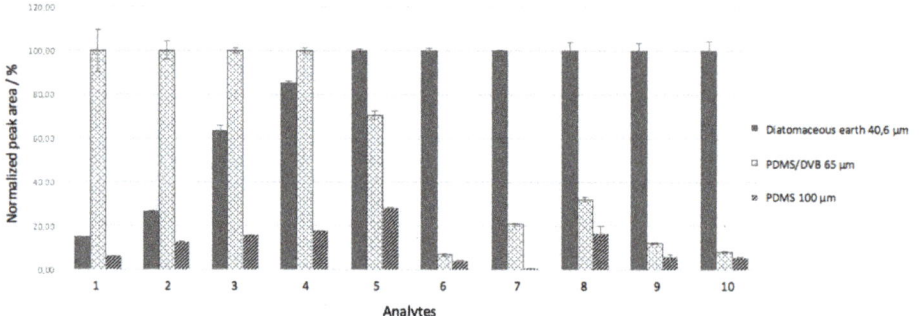

Figure 4. Comparison of extraction efficiencies of the biosorbent fiber, PDMS/DVB and PDMS coatings for determination of PAHs. Analytes: (1) acenaphtlylene; (2) fluorene; (3) phananthrene; (4) anthracene; (5) pyrene; (6) benzo[a]anthracene; (7) chrysene; (8) benzo[b]fluoranthene; (9) benzo[k]fluoranthene; and (10) benzo[a]purene.

It can be observed in Figure 4 that the extraction with the PDMS/DBV coating showed good performance for acenaphthylene, fluorene, phenanthrene, and anthracene, but for the other analytes the results were not as promising. The PDMS fiber gave values below 20%, except for pyrene, and it is not efficient for this application. Taking this into account, the proposed fiber demonstrated very satisfactory performance for PAH extraction when compared to the commercial fibers, with the exception of acenaphthylene, fluorine, and phenanthrene.

In addition, reproducibility studies using two diatomaceous earth fibers were performed and the results showed no significant variation (data not shown). The repeatability obtained with the fibers was estimated comparing the results of the first extraction with those obtained after 115 extractions using the same fiber (data not shown). It was verified that there was no significant loss of extraction efficiency, confirming that the fiber produced with the biosorbent material can be used at least 115 times.

These data demonstrate the high potential for diatomaceous earth fiber as a sorbent candidate for SPME. Moreover, diatomaceous earth is biodegradable, natural, and renewable. In addition, its chemical composition provides numerous possibilities of chemical interaction with a wide range of compounds. Diatomaceous earth has a microporous structure, which facilitates the extraction of the analytes through a physical process (adsorption mechanism).

3.4. Validation Parameters

Table 2 presents some analytical figures of merit obtained in this study. The linear coefficient of determination (R^2) values were >0.95, which indicates a good linear fit. The LOD and LOQ values were satisfactory based on those obtained in other studies.

Table 2. The linear range, linear equation, linearity, and limits of detection and quantification for the method developed using diatomaceous earth coating.

Analyte	LOD ($\mu g\ L^{-1}$)	LOQ ($\mu g\ L^{-1}$)	Linear Range ($\mu g\ L^{-1}$)	Linear Equation	R
Acenaphthylene	0.16	0.49	0.49–25	y = 66,956x − 30,445	0.9890
Fluorene	0.17	0.50	0.50–25	y = 87,809x − 54,517	0.9911
Phenanthrene	0.14	0.42	0.42–25	y = 326,565x − 245,240	0.9777
Anthracene	0.11	0.33	0.33–25	y = 364,057x − 339,574	0.9598
Pyrene	0.15	0.50	0.50–25	y = 979,497x − 935,649	0.9914
benzo[a]anthracene	0.03	0.10	0.10–25	y = 506,040x − 544,597	0.9832
Chrysene	0.14	0.42	0.42–25	y = 691,902x − 796,526	0.9592
benzo[b]fluoranthene	0.06	0.17	0.17–25	y = 158,587x − 50,481	0.9990
benzo[k]fluoranthene	0.11	0.33	0.33–25	y = 431,634x − 806,517	0.9848
benzo[a]pyrene	0.15	0.46	0.46–25	y = 295,450x − 567,387	0.9667

Precision was evaluated in terms of intra-day repeatability (*n* = 3) and inter-day reproducibility (*n* = 9) using samples spiked at the lowest level for each analyte. The results obtained are shown in Table 3. It can be observed that the intra-day and inter-day precision for diatomaceous earth fiber presented values of RSD <15% and <17%, respectively. Relative recovery showed results between 83% and 100%, confirming the accuracy of the method.

Table 3. Relative recovery of analytes and precision (inter- and intra-day) for the extraction of PAHs from spiked river water samples.

Analyte	Spiked Concentration ($\mu g\ L^{-1}$)	Relative Recovery (%) (*n* = 3)	RSD, Intra-Day (%) (*n* = 3)	RSD, Inter-Day (%) (*n* = 3)
acenaphthylene	0.5	100	5	10
fluorene	0.5	83	15	10
phenanthrene	0.5	97	10	13
anthracene	0.5	93	13	3
pyrene	0.5	92	2	6
benzo[a]anthracene	0.5	94	2	6
chrysene	0.5	96	2	6
benzo[b]fluoranthene	0.5	90	15	17
benzo[k]fluoranthene	0.5	97	15	17
benzo[a]pyrene	0.5	93	7	17

The selectivity of the proposed method was confirmed by the absence of peaks in the retention time of the target analytes when chromatograms of the extract were obtained from the river water sample without the addition of the analytes. The only exceptions were pyrene, chrysene, and benzo[a]pyrene, but these peaks were not quantifiable. Figure 5 shows the chromatograms obtained for samples of spiked river water (10 $\mu g\ L^{-1}$) and non-spiked river water.

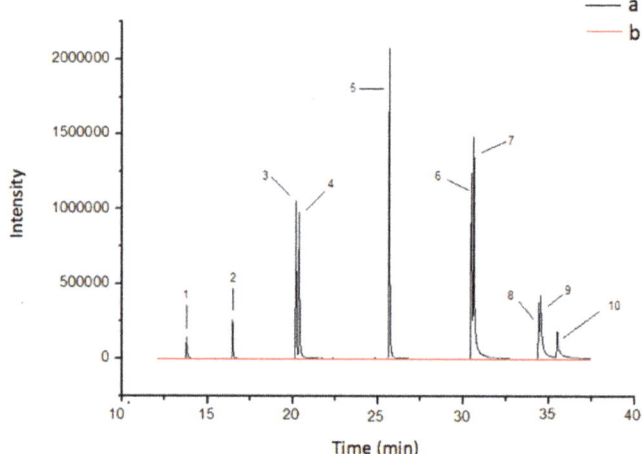

Figure 5. Chromatograms (GC-MS) obtained from a river water sample spiked at 10 μg L^{-1} (**a**) and non-spiked river water sample (**b**). Elution order: (1) acenaphtlylene; (2) fluorene; (3) phananthrene; (4) anthracene; (5) pyrene; (6) benzo[a]anthracene; (7) chrysene; (8) benzo[b]fluoranthene; (9) benzo[k]fluoranthene; and (10) benzo[a]pyrene.

4. Conclusions

In this study, the use of a recycled diatomaceous earth as extractive phase for SPME fiber demonstrated suitable results in comparison to widely used commercial fibers. The production of the biosorbent fiber is simple and the fibers can be reused several times. The separation and detection of the analytes by GC-MS is effective and enables the determination of PAHs in accordance with current Brazilian legislation. The proposed method using the biosorbent achieved good results of parameters of merit. The method is of low cost, because the natural sorbent can be reused in numerous extractions and is widely applicable because the material is easily obtainable.

Author Contributions: All of the authors participated in the same proportion.

Funding: Conselho Nacional de Desenvolvimento Científico e Tecnológico (CNPq), process number 303892/2014-5. Coordenação de Aperfeiçoamento de Pessoal de Nível Superior—Brasil (CAPES) —Finance Code 001.

Acknowledgments: The authors are grateful to the Brazilian governmental agency "Conselho Nacional de Desenvolvimento Científico e Tecnológico (CNPq) and Coordenação de Aperfeiçoamento de Pessoal de Nível Superior" for the financial support which made this research possible.

Conflicts of Interest: The authors declare no conflict of interest.

References

1. Benson, R.; Conerly, O.D.; Sander, W.; Batt, A.L.; Boone, J.S.; Furlong, E.T.; Glassmeyer, S.T.; Kolpin, D.W.; Mash, H.E.; Shenck, K.M.; et al. Human health screening and public health significance of contaminants of emerging concern detected in public water supplies. *Sci. Total Environ.* **2017**, *579*, 1643–1648. [CrossRef] [PubMed]
2. Pal, A.; He, Y.; Jakel, M.; Reinhard, M.; Gin, K.Y. Emerging contaminants of public health significance as water quality indicator compounds in the urban water cycle. *Environ. Int.* **2014**, *71*, 46–62. [CrossRef] [PubMed]
3. Heleno, F.F.; Lima, A.C. Evaluation of analytical methods for BTEX analysis in water using extraction by headspace (HS) and solid phase microextraction (SPME). *Quim. Nova* **2010**, *33*, 329–336. [CrossRef]
4. Hong, W.F.; Jia, H.; Li, Y.F. Polycyclic aromatic hydrocarbons (PAHs) and alkylated PAHs in the coastal seawater, surface sediment and oyster from Dalian, Northeast China. *Ecotoxicol. Environ. Saf.* **2016**, *128*, 11–20. [CrossRef] [PubMed]

5. Slezakva, K.; Castro, D.; Delerue-Matos, C. Impact of vehicular traffic emissions on particulate-bound PAHs: Levels and associated health risks. *Atmos. Res.* **2013**, *127*, 141–147. [CrossRef]
6. Cristale, J.; Silva, F.S.; Zocolo, G.J.; Marchi, M.R.R. Influence of sugarcane burning on indoor/outdoor PAH air pollution in Brazil. *Environ. Pollut.* **2012**, *169*, 210–216. [CrossRef] [PubMed]
7. Dat, N.D.; Chang, M.B. Review on characteristics of PAHs in atmosphere, anthropogenic sources and control technologies. *Sci. Total Environ.* **2017**, *31*, 682–693. [CrossRef] [PubMed]
8. Siritham, C.; Thammakhet-Buranacha, C. A preconcentrator-separator two-in-one online system for polycyclic aromatic hydrocarbons analysis. *Talanta* **2017**, *15*, 573–582. [CrossRef] [PubMed]
9. Li, Z.; Ma, R.; Bai, S.; Wang, C.; Wang, Z. A solid phase microextraction fiber coated with graphene-poly9ethylene glycol) composite for the extraction of volatile aromatic compounds from water samples. *Talanta* **2014**, *119*, 498–504. [CrossRef] [PubMed]
10. Laopongsit, W.; Srzednicki, G.; Craske, J. Preliminary study of solid phase micro-extraction (SPME) as a method for detecting insect infestation in wheat grain. *J. Stored Prod. Res.* **2014**, *59*, 88–95. [CrossRef]
11. Lord, H.; Pawliszyn, J. Evolution of solid-phase microextraction technology. *J. Chromatogr. A* **2000**, *885*, 153–193. [CrossRef]
12. Dias, A.N.; Simão, V.; Merib, J.; Carasek, E. Cork as a new (green) coating for solid-phase microextraction: Determination of polycyclic aromatic hydrocarbons in water samples by gas chromatography-mass spectrometry. *Anal. Chim. Acta* **2013**, *772*, 33–39. [CrossRef] [PubMed]
13. Carasek, E.; Merib, J. Membrane-based microextraction techniques in analytical chemistry: A review. *Anal. Chim. Acta* **2015**, *23*, 8–25. [CrossRef] [PubMed]
14. Pawliszyn, J. *Handbook of Solid Phase Microextraction*; Chem. Ind. Press: Beijing, China, 2009.
15. Tsao, Y.U.; Wang, Y.C.; Wu, S.F.; Ding, W.H. Microwave-assisted headspace solid-phase microextraction for the rapid determination of organophosphate esters in aqueous samples by gas chromatography-mass spectrometry. *Talanta* **2011**, *84*, 406–410. [CrossRef] [PubMed]
16. Ahmadi, M.; Elmongy, H.; Madrakian, T.; Abdel-Rehim, M. Nanomaterials as sorbents for sample preparation in bioanalysis: A review. *Anal. Chim. Acta* **2017**, *15*, 1–21. [CrossRef] [PubMed]
17. Do Carmo, S.; Merib, J.; Dias, A.N.; Stolberg, J.; Budziak, D.; Carasek, E. A low-cost biosorbent-based coating for the highly sensitive determination of organochlorine pesticides by solid-phase microextraction and gas chromatography-electron capture detection. *J. Chromatogr. A* **2017**, *1525*, 23–31. [CrossRef] [PubMed]
18. Suterio, N.G.; do Carmo, S.N.; Budziak, D.; Merib, J.; Carasek, E. Use of a Natural Sorbent as Alternative Solid-Phase Microextraction Coating for the Determination of Polycyclic Aromatic Hydrocarbons in Water Samples by Gas Chromatography-Mass Spectrometry. *J. Braz. Chem. Soc.* **2018**, *29*. [CrossRef]
19. Silveira, C.B.; Goulart, M.R. Methodologies for the reuse of the diatomaceous earth residue, from filtration and clarification of beer. *Quim. Nova* **2011**, *34*, 625–629.
20. Souza, G.P.; Filgueira, M. Characterization of natural diatomaceous composite material. *Ceramica* **2003**, *49*, 40–43. [CrossRef]
21. Othmer, K. *Encyclopedia of Chemical Technology*; Wiley: New York, NY, USA, 1993; p. 108.
22. Menezes, H.C.; Paulo, B.P.; Paiva, M.J.N.; Barcelos, S.M.R.; Macedo, D.F.D.; Cardeal, Z.L. Determination of polycyclic aromatic hydrocarbons in artisanal cachaça by DI-CF-SPME–GC/MS. *Microchem. J.* **2015**, *118*, 272–277. [CrossRef]
23. Aguinaga, N.; Campillo, N.; Vinas, P.; Hernández-Córdoba, M. Determination of 16 polycyclic aromatic hydrocarbons in milk and related products using solid-phase microextraction coupled to gas chromatography–mass spectrometry. *Anal. Chim. Acta* **2007**, *23*, 285–290. [CrossRef] [PubMed]
24. Segura, A.; Sánchez, V.H.; Marqués, S.; Molina, L. Insights in the regulation of the degradation of PAHs in Novosphingobium sp. HR1a and utilization of this regulatory system as a tool for the detection of PAHs. *Sci. Total Environ.* **2017**, *590*, 381–393. [CrossRef] [PubMed]

© 2018 by the authors. Licensee MDPI, Basel, Switzerland. This article is an open access article distributed under the terms and conditions of the Creative Commons Attribution (CC BY) license (http://creativecommons.org/licenses/by/4.0/).

MDPI
St. Alban-Anlage 66
4052 Basel
Switzerland
Tel. +41 61 683 77 34
Fax +41 61 302 89 18
www.mdpi.com

Separations Editorial Office
E-mail: separations@mdpi.com
www.mdpi.com/journal/separations

www.ingramcontent.com/pod-product-compliance
Lightning Source LLC
LaVergne TN
LVHW070046120526
838202LV00101B/648